# Greenhouse Construction

## A COMPLETE MANUAL

ON THE

Building, Heating, Ventilating and Arrangement

OF

# GREENHOUSES

AND THE

Construction of Hotbeds, Frames and Plant Pits

BY

## L. R. TAFT

*Professor of Horticulture and Landscape Gardening, Michigan Agricultural College*

---

## *ILLUSTRATED*

---

NEW YORK
ORANGE JUDD COMPANY
1894

# PREFACE.

In the summer of 1889 the writer erected two forcing houses for the Michigan State Experiment Station. They were designed to be experimental in their construction, and afforded means for a comparative test of various methods of building, glazing and ventilating, and of the relative merits of steam and hot water for greenhouse heating. When the houses had been used one season, a bulletin was issued, in which the construction was described, and the merits or demerits of the methods used were pointed out. During the winter a test of the heating systems was made, and the results were given in the same bulletin. The report was widely distributed, and was copied in full by many horticultural and engineering periodicals, while others gave it favorable notices, which led hundreds of prospective builders of greenhouses, in all parts of the country, to apply for copies, and made a second edition necessary. From nearly every State in the Union came letters, asking advice upon various points in greenhouse construction and heating, all of which indicated, not only that there was a widespread desire for information on these subjects, but that the sources of information were quite limited.

At the request of the publishers, the preparation of this book was undertaken, and the attempt has been made to present the best methods of greenhouse construction.

Although with fifteen years' experience in greenhouse management and a large experience in greenhouse

construction, all of which has been in connection with
the agricultural colleges of various states, where there
was an excellent opportunity of testing the different
wrinkles in construction that have been, from time to
time, brought out, the writer has availed himself of
various opportunities, during the past three years, to
visit the leading floral and vegetable growing establishments of more than a dozen large cities, between Boston
and St. Louis, and has made a careful study of the
methods employed. Many of the leading florists have
submitted their ideas, either in personal interviews or
in correspondence, and from the pages of the *American
Gardening, American Florist, Gardening, American
Agriculturist* and other periodicals, much useful information has been obtained.

Some of the firms that are engaged in the building
and heating of greenhouses have been in business for
many years and have had a wide experience. From
these sources many valuable points have been received, and
it is to their kindness that we are indebted for the illustrations of the exteriors and interiors of some of the
best establishments in the country.

The information here presented has, therefore, come
from a great variety of good sources, and instead of
[illegible]

[remaining text illegible]

# CONTENTS.

V

# LIST OF ILLUSTRATIONS.

vii

# GREENHOUSE CONSTRUCTION.

## CHAPTER I.

### HISTORY OF GREENHOUSES.

It is known that the old Romans were able to secure fresh fruits and vegetables, for their banquets, the year round, by both retarding and accelerating their growth. As an indication of their skill, it is said that they even forced the cucumber. They possessed no elaborate structures for this purpose, but grew them in pits covered with large slabs of talc. Heat was obtained from decomposing manure, and by means of hot air flues. They are believed to have had peach and grape houses, and it is claimed by some, that hot water in bronze pipes was used to warm them.

In modern times the structures have undergone a gradual development, from houses containing no glass whatever, to the forcing house of to-day, which is nearly ninety-five per cent. of glass. The first house of which we have any record, was built by Solomon de Caus, at Heidelberg, Germany, about 1619. It was used to shelter over four hundred orange trees planted in the ground, during the winter, and consisted of wooden shutters placed over a span roof framework, so as to form the walls and roof. It was warmed by means of four large furnaces, and ventilated by opening small shutters in the sides and roof. In the spring the framework was taken down. This structure, in size, com-

1

pared well with the greenhouses of to-day, as it was two
hundred and eighty feet long and thirty-two feet wide.
On account of the expense of putting up and taking
down this framework, and of keeping it in repair, it
was replaced by a structure of freestone. This had an
opaque roof, and the openings in the sides were closed
with shutters during the winter. In 1684 Ray describes
a glass house (Fig. 1) used in the Apothecaries' Garden,

FIG. 1.   ENGLISH GREENHOUSE OF 17TH CENTURY.

Chelsea, England, which evidently was quite similar to
the one at Heidelberg, except that it had glass windows
in the side walls; the roof, however, was opaque. It
was not until 1717 that glass roofs were used, and from
that time, for one hundred years, few improvements
were made.

During the first part of the present century considerable attention was given to the slope of the roof, and
in 1815 the hemispherical form was first used. Before
the use of glass for the roof became common, the greenhouses often occupied the first floor of two-story structures, while the second floor was occupied by the gardener as a residence, or was used as a storeroom.

The earlier greenhouses of this country were not
unlike those used in Europe during the eighteenth cen-

tury. In the *American Florist* for Feb. 15, 1887, is the description and figure of what is supposed to be the

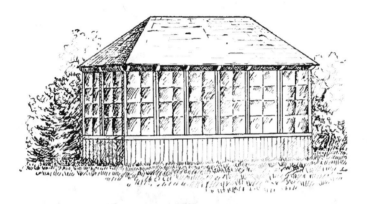

FIG. 2. FIRST AMERICAN GREENHOUSE.

first American greenhouse (Fig. 2), it having been erected in New York, in 1764, for James Beekman. Although the structures were less elaborate, the American builders took up and utilized any improvements in construction and heating that were brought out in Europe.

In *Hovey's Magazine of Horticulture*, for January

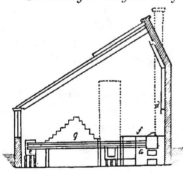

FIG. 3. MODEL GREENHOUSE OF 1835.

1836, is a description of a model greenhouse, erected by Mr. Sweetser, of Cambridgeport, Mass. From Fig 3, in which a cross section is shown, it will be seen that glass was used in the entire south slope of the roof and in the south wall. The north slope of the roof and the north wall were of wood. The heating system combined the flue with hot water. The hot water system consisted of an open copper kettle, or heater (*f*), from the top of which a four-inch copper pipe passed across the end of the house, and then along the opposite side, to a large copper reservoir (*e*); the return pipe was located on a level, just beneath the flow, entering the boiler near the bottom. The flue was carried to one side until it reached the walk (*c*), and then ran under this to the other end of the house, where it was connected with the chimney (*d*).

In the West, greenhouse construction was more backward, and yet, as early as 1836, a Mr. Thomas,

FIG. 4.　FIRST CHICAGO GREENHOUSE.

according to the *American Florist*, erected one in Chicago, of which an illustration is shown in Fig. 4. As will be seen from the engraving, the three-quarter span houses had even then come into use, although the entire north slope, and half of the south slope of the roof, were of wood.

Previous to 1850 there were comparatively few greenhouses in the country, and, naturally, there were no extensive builders. Among the first to engage in the business was Frederic A. Lord, who erected his first houses in Buffalo, in 1855. In 1870 he removed to Irvington, and in 1872 entered into partnership with W. A. Burnham, under the firm name of Lord & Burnham. In 1883 the firm of Lord & Burnham Co. was incorporated. The earlier houses erected by this firm were, for the most part, in the curvilinear style, which, in a slightly modified form, is still used by them for large conservatories.

In 1888 the firms of Hitchings & Co., and of Thos. W. Weathered's Sons, both of New York City, who for many years had been engaged in greenhouse heating, added departments for greenhouse construction. John C. Moninger, of Chicago, is one of the best known builders in the West. The systems of construction used by the three first mentioned firms are much alike, their better houses being put up with iron sills, posts, rafters, purlins, etc., while the sash bars are of cypress, although many of their commercial establishments have no iron in their construction.

The puttyless glazing systems have been but little used, except in large conservatories. The principal firms controlling them are the Plenty Horticultural Works of New York City, and A. Edgecomb Rendle & Co. of Philadelphia and Chicago, each of whom have been in business for some ten years, and have erected a number of large establishments. They also use the wooden sash bars and putty glazing. Nearly every large city has one or more dealers in structural iron work, who have taken up greenhouse construction. Most of them use galvanized iron, with or without a steel core, for sash bars. The use of iron for the rafters and sash bars of fixed roofs, has been quite general in England for eighty years, as its permanency is there thought to more than counterbalance the extra expense, breakage of glass, loss of heat, drip and leakage, with this system, as compared with wooden supports. In this country the winters are much more severe, and, the conditions being less favorable for iron roofs, their use is not regarded with favor by commercial florists.

We have no description of the furnaces used by Dr. Caus, in his orangery, but Evelyn tells us that the Chelsea greenhouse was heated by an open charcoal fire built in a hole in the ground. Later on, a chimney was carried through the greenhouse, and this developed into the greenhouse flue, which is still in use. Although steam was tried for heating greenhouses, in the last quarter of the eighteenth century, it was not much used until about 1816, when, for twenty years, it was in high favor, but was superseded by hot water, which, in turn, has, during the last few years, been crowded out, in large plants, by steam.

## CHAPTER II.

### DIFFERENT FORMS OF GREENHOUSES.

While the various glass structures are generally distinguished according to their uses, as rose houses, palm houses, stove houses, graperies, etc., for our present purpose it will be well to first consider them from the builders' standpoint, as lean-to, span roof, three-quarter span, and curvilinear houses. These names have been applied from the various shapes that may be given to the houses. While any of these forms of houses *may* be used for all purposes, each one of them is particularly adapted for the growing of certain plants, and as they each have their special advantages and disadvantages, they should have careful consideration.

### SPAN ROOF HOUSES.

The form of glass structure which has come to be known as the span roof is, more properly, the "even span," as the lean-to may be considered a "half span" house, while we also have "two-third" and "three-quarter span" houses. The typical "even span" house is generally from nine to twelve, or from eighteen to twenty feet wide, with side walls from four to five feet high. The two slopes of the roof are of the same extent, and are arranged at the same angle, usually between thirty and thirty-five degrees, which will bring the ridge, in a house twenty feet wide, about ten feet above the walk, in a house with walls four feet high, and the roof at an angle of thirty degrees, and eleven feet high, when it has a slope of thirty-five degrees.

In a house of this size, it is desirable to have, at least, two rows of ventilating sash, which may be on either side of the ridge, or, if three rows are used, one

FIG. 5. EVEN SPAN GREENHOUSES.

may be located at the ridge and the others in the side walls. The amount of ventilation desirable will, of course, be determined largely by the plants to be grown in the house.

Although less simple in construction than the lean-to, they have a far greater variety of uses, and are much more frequently erected. In fact, nearly all the houses constructed for commercial purposes, prior to 1885, were of what is known as the even span style (Fig. 5).

For ordinary growing houses for a commercial florist, this form is as good as can be secured, although for forcing houses the three-quarter span is preferable. One of the special advantages of these houses is that they may be run at almost any direction that the location may necessitate. With this form of roof, the benches can be placed at the same height, and the plants will still be near the glass, while in other forms of roof, with the same pitch, the ridge will be much higher.

The even span houses are usually run north and south, as this not only brings the plants, on both sides of the houses, into full sunshine during a part of the day, but better than any other direction, or any other kind of a house, provides for a perfect distribution of the rays of light and heat upon all sides of the plants. For many purposes the east and west arrangement, with one side facing the south, is preferable, as, during the four hours of the day when the sun's rays are most powerful, they strike at right angles to the glass, and are but little obstructed by the sash bar ; while, were the houses running north and south, more than half the rays would be cut off between eleven and one o'clock, and, as this part of the day is particularly valuable in the forcing house, this arrangement is preferable for them. The beneficial effects, however, will be confined to about two-thirds of the house, on the side towards the sun, while the other side will have much less sun than were it in a house running north and south. If designed as growing houses, this might not be objectionable, as the north side could be used for ferns, violets, or for plants at rest, which do fully as well in partial shade.

The fact that the north third of the house is of little value for forcing purposes, led, in part, to the construction of the first forms of two-third and three-quarter span houses, which, so far as the slope of the roof is concerned, did not differ from the even span, the only difference being that the back wall was run up at a point which cut off the north third or fourth of the house. Everything else being equal, the loss of heat from a span roof house will be somewhat greater than from either a lean-to, or uneven span house, especially if it, like the others, runs east and west, on account of its having a greater area of glass upon its north side. In the lean-to there is no glass at all on the north side, while, in the three-quarter span house, the glass area on the north side will only be one-half as great as in the even span.

The even span houses may vary in width, from nine to twenty-four feet outside. For the narrow houses only one walk, situated in the center, with a bench on each side (See Fig. 59), is used. When the walk is two feet wide there will be room for two tables, each three and one-half feet in width. These widths may be increased to four feet for the beds, and two feet six inches for the walk, if necessary, but twelve feet would be the extreme width that could be used with comfort, when a house with a single walk is to be used for most greenhouse crops, especially if they are grown in pots. When two walks are used, the houses would need to be increased to a width of, at least, sixteen feet, and, for some purposes, may be as much as twenty-four feet, which will be as wide as will probably be used under any circumstances. A medium width, however, is preferable, and the greatest economy of space and comfort, in caring for the houses, will be obtained, when the houses are not less than eighteen feet, nor more than twenty, outside measurement.

While houses are often built with walks as narrow as eighteen inches, it is better to allow two feet, in commercial growing houses, and in private houses a width of two and one-half feet for walks, will not be too great. For the side benches, three feet and six inches will be

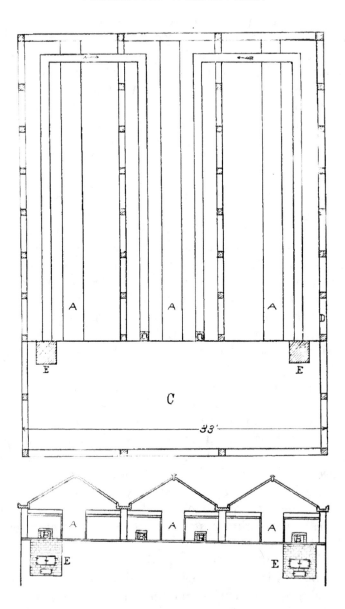

FIG. 6.   GROUND PLAN AND END ELEVATION OF RIDGE
AND FURROW HOUSES.

found a convenient width, although four feet is often used. If the greatest economy of space and convenience of handling the plants is sought, the center bench should be about seven feet in width. They are, however, often made as narrow as six feet, and when large plants are to be grown, which will make a high roof desirable, the width of the house may be increased to take in a bench ten or eleven feet in width.

### RIDGE AND FURROW HOUSES.

The even span form of roof has one advantage that is possessed by no other (except by the short span to the south, or the three-quarter span on a sidehill), as they admit of the ridge and furrow construction, as it is commonly called. This should, however, be distinguished from the ridge and furrow form used in England by Sir Joseph Paxton and others, in which the roof was broken up into a great number of ridges, and furrows run up the main slope of the roof.

The principal gain is due to the fact, as shown in Fig. 6, that when, say five, houses are built in this way, only six walls will be required, and four of these can be of light, cheap construction, instead of the ten well-built walls that would be necessary were the houses built separately. Another advantage, which should not be overlooked, is due to the fact that there will be only one-fifth as much exposed wall surface, and that, when built thus close together, one house on each side will protect the others from the high, cold winds that come from that direction. There will also be a considerable saving in space, which will be worth considerable, especially in cities.

Among the disadvantages of the ridge and furrow style of houses, is the shading of the center houses during the morning and afternoon, by those on either side, by which, especially in the case of wide, high houses, much light and heat is shut off; also the fact that when houses are built in this way, side light and side ventilation cannot be secured. While this is not even desirable for some crops, for others it is quite necessary, and whether crops are to be selected that are adapted to the houses, or houses are to be erected that are to be suited to the growing of certain crops, this should be understood.

It may be laid down as a rule that, aside from the economy of erection, heating, etc., better results will be obtained from wide houses, if they are built with intervals of, at least, fifteen feet between them; but when the erection of the extra walls, and the increase in fuel and land are considered, for the ordinary florist, even span houses, of a width of twenty feet or less, should be erected upon this plan, unless other special reasons might exist. In sections where the snowfall is heavy, the gutters will become filled, and, as the snow cannot slide from the roof, with long houses, unless they are narrow, and only built with three houses in a section, to allow of the snow being thrown over the roofs of the side houses, this will be a serious objection to the plan. For the growing of small bedding plants, mignonette, heliotrope, carnations, and for propagating houses, this form of construction, with houses twelve feet wide, will be quite satisfactory. Of course, any plants can be grown in them, but a wider house seems preferable for roses, carnations, lettuce, and for most forcing crops, particularly as the amount of air enclosed is greater in proportion to the amount of exposed glass surface, on which account the temperature can be easier regulated, and drafts of air prevented.

### THE LEAN-TO HOUSE.

When it is desirable that the first cost shall be as small as possible, and if the expense for fuel, rather than the crops grown in the house, is considered, the lean-to form, particularly if the structure is to be a small one, will be found of value. An idea of the shape of the house, and the reason for the name, can be obtained from Fig. 7. If the house can be built against the south wall of a building, or against a steep sidehill, these will be additional reasons, as affecting the cost of erection and heating, for using this form of construction. On the other hand, this shape for a greenhouse has, per-

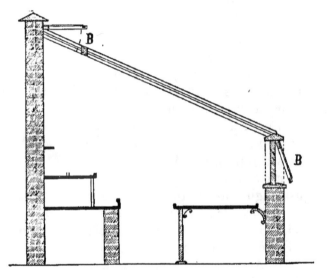

FIG. 7. LEAN-TO HOUSE (*Cross Section*).

haps, more and greater objections than any other. One serious fault is that for three hours in the forenoon, and an equal period in the afternoon, the plants get little or no direct sunlight; another objection is that the light that they do get, coming all from the south side, is unequally distributed upon the plants, and the leaves are all turned in that direction, thus giving the plants an uneven appearance.

As grape or peach houses, the lean-to construction answers very well, and, where one has a wall that can be utilized, the expense for building and heating will be very small. A lean-to, with its roof sloping to the north, answers very well as a propagating house. One of the simplest ways of building one is to place it against the north wall of a three-quarter span house, or by building an even span house twenty-five feet wide, and cutting off six and one-half feet on the north side, thus forming, on one side, what is known as a north side propagating house (See Fig. 61), and on the other a three-quarter span forcing house. As a small house for an amateur, quite satisfactory results can be obtained from a lean-to, but a span roof house is to be preferred.

For the forcing of vegetables, the growers of lettuce at Arlington, Mass., and vicinity, use wide houses constructed on the lean-to plan (See Fig. 83), and they give excellent satisfaction. As a rule, lean-to houses are built with a wall from four to six feet high, and with a roof of a width in proportion to the width of the house; but they are sometimes built quite narrow, with a low wall, just high enough to allow of bottom ventilation, from which the side sash rises at an angle of from forty-five to sixty degrees, to a height of eight or nine feet, with a narrow ventilator connecting the top with the back wall. A good idea of the form of this house can be obtained from Fig. 97. The principal use of a narrow lean-to of this kind, would be as a cold grape or peach house.

In a general way, the construction of a lean-to house would be the same as of half of a span roof house, and, so far as the building of these houses is concerned, they will be treated under the same headings, and will receive no further consideration as distinct houses.

### SIDEHILL HOUSES.

A modified form of lean-to, which combines its advantages with those of the three-quarter span house,

is sometimes known as the sidehill house. W. C.
Strong, of Massachusetts, erected a house of this kind at
Brighton, and was well pleased with it. Other smaller
houses have since been erected, and, for vegetable forcing, have given excellent satisfaction. A good idea of
the construction of the houses can be obtained from
Fig. 8. They should be located upon a hillside which
has a slope towards the south of about twenty-five degrees. Each section should consist of a lean-to structure
of any desired width, from ten to twenty-five feet. The
south wall is built the same as for any greenhouse, and,
for a structure fifteen feet wide, posts should be placed

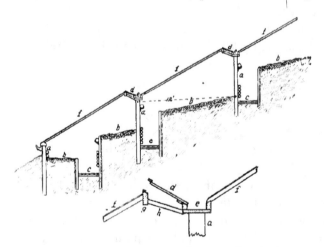

FIG. 8. SIDEHILL HOUSES (*Section*).

in the ground, as at *a*, for the north wall; a gutter
should be placed upon them, and this will answer for
the south gutter of the adjoining house. The sash bars
should be laid at the same angle as the slope of the hill,
against a ridge. *g*, which should be about two by four
inches, as should the sides of the gutters, *e*.

The ridge is supported by braces about two by four
inches, which are placed at intervals of two and one-half

feet, as is shown at *h*. The ventilators, the construction of which is shown at *d*, are of wood, and will be found convenient to walk upon in removing the snow and making repairs, otherwise they could be of glass, if preferred. The benches may be arranged as is most convenient, the method shown in Fig. 8 being an excellent one. The heating pipes may be arranged along the sides of the walks, but should be so distributed that the lower houses will have their share of the heat.

In Europe, houses of this form are very commonly used, and vegetables of all kinds are grown out of season, in much the same way as in the open air. Hundreds and thousands of acres are thus covered with glass, and the profits of a quarter of an acre are sometimes more than from the best hundred acres used in general farming.

## CHAPTER III.

### THREE-QUARTER SPAN HOUSES.

As previously stated, the first form of three-quarter span house was the same as three-quarters of an even span structure, but the shape of the roof has been somewhat modified, so that the plants will be nearer the glass. The cost of building these houses is about the same as for an even span, but owing to the fact that the north wall is from six to eight feet high, there will be less loss of heat from the north side of the roof, and the south pitch of the roof will take in more of the light and heat rays, than would be the case with a span roof house.

The three-quarter span houses may be likened to a lean-to house with the peak of the roof cut off. In the lean-to the heat tends to rise into the angle of the roof, and hence is not evenly distributed, but in the three-quarter and even span houses there is less trouble from this. The three-quarter span houses always run east and west, and the north slope of the roof allows the light to fall on the plants from all sides, so that the growth of the plants will be stronger and more symmetrical. It is the south slope that is principally relied upon to trap the light and heat of the sun, and the angle at which the glass is arranged is that which will be nearest at right angles to the sun's rays during the winter months.

This form of house is particularly adapted to the forcing of roses, and of all other plants that need a maximum amount of light for their development. In Fig. 9

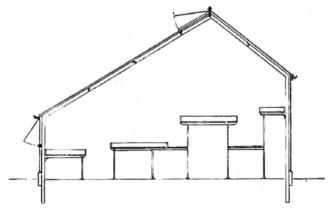

FIG. 9. THREE-QUARTER SPAN HOUSE (*Section*).

will be seen the usual form of forcing house of the three-quarter span style. For adapting it to different crops, the height of the walls, the slope and length of the sash bars, and the width and height of the benches, can be varied at pleasure. As a general rule, the three-quarter span houses are from sixteen to twenty feet wide; the south wall is from four to five feet high, and the north one from six to eight feet. The south pitch of the roof

2

varies from twenty-six to thirty-five degrees, and the north one from thirty-five to sixty-five degrees.

The side benches are each about three feet wide, and are placed about one and one-half feet below the plates. The center bench may be single (Fig. 63), with a slope to the south, or double, as shown in Fig. 9, with a narrow walk between the two parts. This style of house is also largely used for lettuce forcing, and for this purpose the width is sometimes increased to thirty-five feet.

## CURVILINEAR ROOFS.

In this construction the sash bars are more or less curved, with the idea that, at all times of the day, some of the glass will be at right angles to the sun's rays. This, of course, is secured, but the result of the curved sash bars is to decrease the angle at which the rays strike a majority of the panes, so that, after all, the curvilinear construction is an injury, rather than a benefit. The old style of curved roof had the sash bars leaving the plate in nearly a vertical direction, and with most of the curve in the lower third of the roof. As a result, the upper half of the roof approached the horizontal, and made a very small angle with the sun's rays, especially during the winter. The present form of curvilinear roof has a more regular curve, and, as shown in Fig. 10, is less objectionable. Whatever the material used, the cost of the framework for a curvilinear house is considerably more than for a straight roofed house. If, for glazing the roof, glass bent to the proper angle is used, the cost will be much more than for straight glass. Ordinary sheet glass can, of course, be used upon curvilinear roofs, but especially upon the old form of roof, comparatively short panes must be used.

To many persons the curve is a "line of beauty," and a curvilinear house has a more ornate and finished

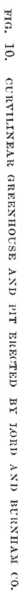

FIG. 10. CURVILINEAR GREENHOUSE AND PIT ERECTED BY LORD AND BURNHAM CO.

appearance than one with straight sash bars and, in private and public parks, where the increased cost is not considered an objection, and where the houses would be an ornamental feature of the landscape, curvilinear houses have their place. This form of roof is also quite desirable for large conservatories, although a roof made with straight sash bars can be so broken up as to relieve it of any barn-like appearance. The curvilinear construction can be used in lean-to, even span, or three-quarter span houses, but for the reasons given is not particularly desirable in any form of low, narrow houses, and, in fact, it is generally admitted that better plants can be grown in houses with straight sash bars.

Some twenty years ago the curved construction was in very common use in England, but the general verdict seems to be expressed by a writer,* who says, "Taken as a whole, circular work may, in a few exceptional instances, be introduced to obtain an architectural result, or in molding the lines of a large winter garden or magnificent palm house, but for ordinary growing purposes, we may consider curvilinear roofs not so suitable as those composed of straight lines." As a result of this belief, the curved roofs are no longer in favor, and few such are being erected to-day.

* Fawkes. Horticultural Buildings, P. 54.

# CHAPTER IV.

## LOCATION AND ARRANGEMENT.

When erected in connection with some other building, the aspect and slope cannot always be regulated; but, if possible, greenhouses for most purposes should be on the south side, so that no rays from either east or west will be cut off. For a lean-to or a three-quarter span house the wall or building against which they are erected should run east and west, and an even span house should, in this case, run north and south, with its north end against the other structure.

For the location of detached houses, if thorough drainage can be secured, a level spot is not objectionable; while, if it is at the top of a south and westerly slope, all the better, as there the sun can get in extra hours at both ends of the day. In case the land on the most available site is not level, it should be graded, in case it can be done without too great expense. A slope of perhaps one or two inches in fifty feet, to carry off the water from the gutters, is not objectionable, and, while it is preferable that each house should be practically level, if the land selected cannot be readily graded so as to bring all of the houses upon the same level, there will be no serious objection to having the houses ranged, one above the other, in regular tiers. For sidehill houses a decided slope is necessary. In locating the houses, means of thorough drainage, particularly for the boiler room, should be the first desideratum. In arranging a group of houses, the width and height of the different structures, and the shape of the roofs, will have much

21

to do in determining their exact location. Unless
arranged in ridge and furrow style, a space of twelve or
fifteen feet between the houses is desirable. The lean-to
and three-quarter span houses may be placed in parallel
lines running east and west, and the even span houses
may run in either direction. Besides having them so
located as not to shade one another, to prevent side ventilation, or, if desirable, driving between them with a
horse and cart, they should be as near together as is possible, in order to save land, and for convenience and
economy in heating and operating the houses.

The convenience of arrangement assists, to a wonderful extent, in the performance of the greenhouse
work. The potting and workrooms should be centrally
located, well lighted, and in every way convenient for
the work, and in commercial establishments the packing-
room should be so situated as to facilitate getting up the
orders. In retail establishments, when the salesroom is
in connection with the greenhouse, it should be conven-
iently located for the customers, and should be fitted up
with counter, glass show cases, refrigerator and other
necessary furniture, Fig. 76. If properly arranged, with
the wire designs upon wall hooks, the baskets and similar supplies in glass cases, in fact, with a place for everything, and everything in its place, the salesroom will be
attractive to customers and visitors, while its convenience, and the arrangements for preserving the flowers
and supplies, will soon repay all expense. We believe
the above equipment to be almost a necessity in a properly conducted business, and if there is a large retail
trade at the greenhouse, some attempt at decoration,
both in the salesroom, and in one house to be used in
whole or in part as a showroom, cannot fail to attract
visitors, and this will increase the trade.

In locating the various workrooms for a large establishment, it is well to have them in the center, with the

houses running out from both sides, east and west. A similar arrangement for the heating plant is also desirable; thus, rather than have the boiler room at one end of a long range of houses, the boiler house could be placed in the center, and houses of half the length arranged on each side, and better results obtained. A very convenient arrangement for the heating plant is shown in an engraving of F. R. Pierson's range of rose houses, in Chapter Twenty-two, in which four houses, each one hundred and fifty feet long, are supplied from a boiler house so located that the extreme ends of the houses are but little more than one hundred and fifty feet away, instead of being over three hundred feet, as would be the case were the boilers located at the end of the range. Of course, with houses of one hundred to one hundred and fifty feet, such an arrangement would not be desirable. In many establishments it would be convenient to widen the connecting passage-way, and use it for potting, packing and like purposes. For a rose forcing house, the potting and packing rooms need not be as large nor as centrally located as for an ordinary commercial establishment. In large private establishments, the palm house is generally the central figure, around which the others are grouped. For a small range the one shown in Fig. 100 is well planned, while that in Fig. 106 has as many merits as a large one.

# CHAPTER V.

## GREENHOUSE WALLS.

In erecting greenhouses, too little attention is usually paid to the construction of the walls. Not only should their durability be secured, but the heaving by the frost, and the lateral pressure of the roof, should be guarded against. Owing to a lack of foresight regarding some of these points, one seldom sees a greenhouse with five years' service that is in a satisfactory condition. Greenhouse walls are constructed of wood, brick, stone or grout, or of a combination of two of these materials. Each of these methods of construction has its advocates, but each of them has some disadvantages.

### MASONRY WALLS.

The use of stone or grout (cement, sand and cobblestones), for the construction of the foundation of brick walls, is very common, and, as they make a durable wall, would, no doubt, be largely used for the walls up to the plates, were it not that they are rapid conductors of heat. In small greenhouses, where the grade can be carried up to the plate, so that none of the wall is exposed to the outside air, they make excellent walls.

The excavation should be to a depth of three feet below the proposed outside grade level, and of a width to admit of a fifteen or eighteen inch footing course. This should occupy the trench up to the level of the interior of the house, at any rate, and even if brick or other material is used for the upper part of the wall, may extend to the level of the ground outside, which is

24

often from two to five feet above that of the interior. Stone conducts heat quite rapidly, and for that reason will not be desirable as a wall above ground, unless made very thick. This objection does not hold to the same extent with grout, and where small stones can be readily obtained, it makes a cheap and very durable wall. For a house not over twenty-five feet wide, and when less than five feet in height, a wall of grout twelve inches thick will answer. This should rest on an eighteen inch footing course of the same material. The materials required are, stones from two to four inches in diameter, gravel, and water lime, of Louisville or a similar brand.

In making the wall, a box of the desired width is made by driving stakes along the line of the wall, on each side, and setting up twelve inch planks for the sides of the box. In this a layer of stones is placed, which should be packed in carefully, and kept, at least, one-half inch away from the planks. The cement is then prepared by thoroughly mixing one part with three parts of gravel, and then adding water enough to thoroughly moisten it. The best results are obtained, if it is of about the same consistency as ordinary lime mortar. The water should not be added until the cement has been mixed with the gravel. A layer of cement from two to three inches thick, over the stones, will be sufficient; this should be well tamped down, filling all of the space between the stones. Another layer of stones and cement can then be added, and the process repeated until the box is filled,

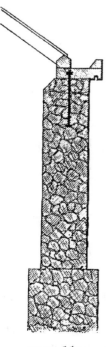

FIG. 11.

GROUT WALL.

requiring about three layers.  One wall of the house can
be built at a time, although if planks are at hand it will
be well to allow one wall to set while a course is being
put in on another.  After the grout has been setting for
five or six hours, the planks can be raised their own
width, and the box will thus be prepared for another
course.  In this way a wall of any desired height can be

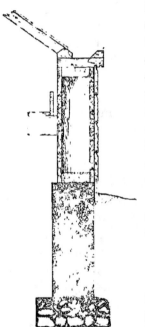

built, which will be found quite
durable and in every way satisfactory.  The appearance of the wall
can be improved if, after the last
course has been put on, the exposed surface is given a thin coat
of Portland cement mortar.  If
desired, the surface can be laid off
into squares, resembling blocks of
sandstone.  The appearance of a
wall built entirely of grout is
shown in Fig. 11, while Fig. 12
shows a wall half grout and half
wood.

### BRICK WALLS.

Unless the very best materials
are used in their construction, the
greenhouse walls constructed of
brick will be comparatively short-

FIG. 12.   GROUT AND  lived, as the combined action of
WOODEN WALL.      moisture and frost will disinte-
grate the mortar, and cause the outer tier of bricks to
crumble.  Hard burned bricks should be selected, and
the best Louisville, or, better yet, Portland cement mortar, should be used.  Whatever the thickness of the
wall, there should be, at least, one air space, to prevent
radiation of heat.  This will also tend to render the wall
more durable, by preventing the capillary passage of the

moisture. For all low walls, two tiers of brick, with a one-inch air space, making a nine-inch wall, will answer ; these should be firmly tied together every fourth course vertically, and every three or four bricks along the walls, Fig. 13. A post once in eight feet will strengthen the wall, and prevent the plates from spreading. For heavy or wide structures, or if the wall is high, a third tier of bricks on the inside, one-half, or, perhaps, two-thirds the height of the wall, will serve to strengthen it. Fig. 14 shows the construction of such a wall, as used with an iron sill.

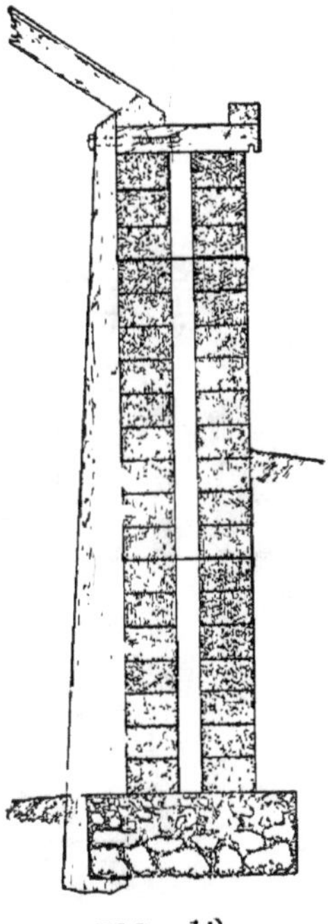

### WOODEN WALLS.

Probably nine-tenths of the present greenhouses are constructed with what might be called post and board walls. In their erection, the posts used should be of some durable material, such as red cedar, locust or cypress, and the size should vary from four by four inches for low walls and narrow houses, to six by six inches for high walls and wide houses. The posts should be seven to eight feet in length, except on the back side of three - quarter span houses, where a length of ten or twelve feet will be necessary, which will allow of their being set three feet in the ground for the front wall, and four feet for the rear one. The posts should be placed in a straight line, about four feet apart, and unless the ground is quite firm and solid, it is well to place a flat stone under the post, and fill up the hole around it with grout. This will not only hold the post firmly in place, but it will

FIG. 13.

BRICK WALL WITH WOODEN SILL.

have a tendency to preserve it. The durability of the posts can also be increased by charring the lower end, and then soaking it in crude petroleum. A coat of coal tar would be better than the petroleum, but it should never be used about a greenhouse, as it will be injurious to the plants.

The posts should then be sheathed upon the outside, for which purpose a fair grade of matched lumber is

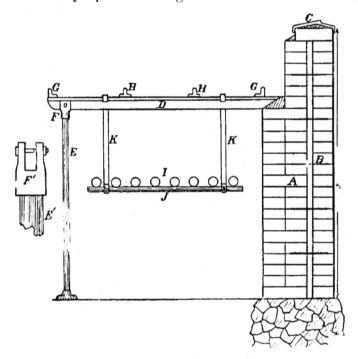

FIG. 14. BRICK WALL WITH IRON SILL.

desirable, although any kind of culled lumber will answer (Fig. 15). It will always pay to cover the sheathing with some kind of heavy building paper, avoiding all brands that contain tar. For the outer covering the novelty or patent siding will be found preferable to ordinary clapboards. For rose and stove houses it may pay, in exposed localities, to ceil up the posts on the inside; but if this is done it will be best not to pack the enclosed space with sawdust or similar material. This course was recommended for many years, but, in practice, it was found that the packing absorbed moisture and caused a rapid decay of the wall. In ceiling up the inside of the posts, tight joints should be made, that will exclude mice; otherwise, the enclosed space may become a harboring place for them, and thus prove a greater injury than benefit. When used for growing roses and other tall plants, that require the bed to be situated at least two feet below the plate, it is well to have a row of sash in each of the side walls (Fig. 16). If houses run east and west, a row along the south side will answer, although one on the north side will be of advantage; in north and south houses the sash should be placed in both sides. In narrow houses they may be fastened permanently, but, if the houses are wide, it will be advisable to have, at least, a part of them on hinges, so that they can be opened if found necessary (Fig. 50).

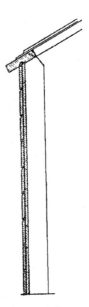

FIG. 15.

WOODEN WALL.

### PLATES AND GUTTERS.

The wall plates may be placed level, on the top of the walls, as in Fig. 12, or they may be at the same angle as the roof. As a rule, two-inch lumber is heavy enough, although if a gutter is desired for catching the roof-water, strips may be nailed to it, as shown in Fig. 12. Another method of arranging the gutter is shown in Fig. 16. Whichever method is chosen, the posts, where wooden walls are used, should all be cut off at the same angle, and the plate securely fastened in place. The arrangement shown in Fig. 16 is the neatest and best, and the same form of plate, without the gutter, can be used when one does not desire the latter. The form illustrated in Fig. 12 will be a cheap and satisfactory method of arranging the gutter, while that shown in Fig. 15 is, perhaps, the cheapest and easiest way of

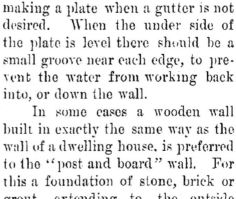

making a plate when a gutter is not desired. When the under side of the plate is level there should be a small groove near each edge, to prevent the water from working back into, or down the wall.

In some cases a wooden wall built in exactly the same way as the wall of a dwelling house, is preferred to the "post and board" wall. For this a foundation of stone, brick or grout, extending to the outside grade line, is necessary, and an excellent plan is to have two, or even three feet of the wall below this level, with a corresponding excavation for the house, necessitating the erection of a wooden wall of the same height above. In this way the

FIG. 16.    WOODEN exposed surface is greatly reduced,
WALL WITH GLASS and a durable and warm wall will be
SIDE.    secured. The sill for this wall
should be two by four-inch scantling, with studding of the same size, placed two feet apart. The sides and top can be arranged in the same manner as when posts are used (Fig. 12). If there is danger of lateral pressure, the sills should be securely anchored to the foundation. This form of wall is principally desirable for narrow houses, and where side ventilation is not needed.

### IRON POSTS AND SILLS.

Although wood is now almost' universally used in the construction of commercial houses, many of the more enterprising florists are employing iron and steel in such portions of the house as do not form a part of the exterior, particularly for posts, sills, purlins and ridge.

One of the simplest and best arrangements of this kind was used by Lord & Burnham Co., in the construction of a range of rose houses for F. R. Pierson, at Scarborough, N. Y. (For views and sections of these houses see Figs. 78 and 79.) The posts and rafters on each side were made from one piece of four by one-half inch bar iron, bent at the gutter-line, so that when the lower end was vertical, to form the post, the other would be at the proper angle for the rafter. The post end was placed in an excavation three feet deep, resting on a flat stone, and the hole was filled up with grout. The upper end

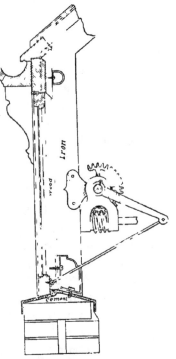

FIG. 17. IRON POST AND SILL, WITH SIDE VENTILATION.

of the rafter was securely bolted to its companion from the other side, by means of an iron bracket. In some cases an iron ridge of the same size of the rafters is used, to which the rafters are fastened by means of angle brackets.

The wall may be constructed in various ways, one of which, illustrated in Fig. 9, will answer well for rose houses. If desired, the entire wall beneath the plate may be of wood, although the glass will generally be found desirable. It will be noticed that the wooden portions of the wall are bolted to the posts by means of small iron lugs (Fig. 17).

The form of post used by Hitchings & Co. is rather more elaborate and ornamental than the above. It consists of a cast iron post base below ground, to which the T iron wall post is bolted. The rafter is then bolted to the top of the wall post by means of an ornamental iron bracket. In the more elaborate conservatories, with a brick or masonry wall, an iron sill, Fig. 17, is used (a similar sill of iron can also be used upon the top of a wooden post if desired), to which the lower end of the rafters is fastened by iron lugs. .

This is the best form of construction now in use, and when ready capital is at hand for the erection of good houses, it will be found most economical in the end. As a second choice, one of the forms of iron posts, with a wall of wood, could be used. Even if the wooden wall does decay, the posts, rafters and purlins will still remain, forming a stiff and firm framework, which would still support the superstructure. As a rule, the board at the bottom will decay first, and if this is so put in that it can be easily taken out and renewed, the wall can be kept in repair for a long series of years at a small expense.

# CHAPTER VI.

The portion of the house to which the most attention is paid, are the strips supporting the glass. There are dozens of patent sash bars, and methods of glazing, and yet the old wooden sash bar is still preferred by the commercial florist, while the sash bars in some of the best modern houses are identical in size and shape with those in use thirty years ago. Although the glazing should be so tight that no water can pass through into the house, there will be more or less condensed moisture on the under side of the glass, and to prevent drip as much as possible, it is well to have them with drip gutters on each side. Some florists, however, prefer not to have them.

Ordinary white pine makes a good sash bar, and, if kept well painted, will be found quite durable. The southern cypress, however, is generally preferred. It is straight grained, rather more durable than white pine under the best of care, and much more so if they are neglected. Cypress is also stronger and stiffer than white pine, and the sash bars can be made rather smaller on that account. For use with lapped glass, the best form of sash bar, if drip gutters are wanted, is shown at Fig. 18, while, if the drip gutters are not desired, a good form is shown in Fig. 19. When glass from fourteen to eighteen inches wide is used, the roof sash bars should be from one and one-eighth by two inches to one and one-fourth by two and one-half inches, according to the distance between the purlins. The rabbets for the glass

3                    33

should be about half an inch deep and five-sixteenths of
an inch wide. The vertical sash bars for the sides and
ends should be about one and one-eighth by one and

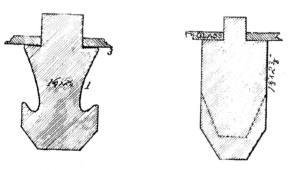

FIG. 18.    SASH BAR WITH    FIG. 19.    PLAIN SASH
DRIP GUTTERS (*Section*).    BAR (*Section*).

seven-eighths inches, the rabbet being of the same size
as for the roof sash bars. For butted glass, whether used
with or without glazing strips, either of the above forms

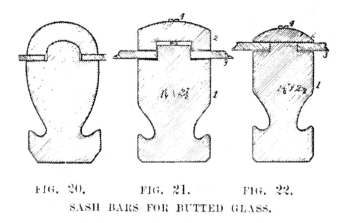

FIG. 20.    FIG. 21.    FIG. 22.
SASH BARS FOR BUTTED GLASS.

(Figs. 18 and 19) may be used. The patterns shown in
Figs. 20, 21 and 22 are, however, preferable for this kind
of glazing. The sash bars (1) there shown are practically
alike, and the difference lies principally in the form of

the caps (2), those shown in Figs. 20 and 21 being, perhaps, preferable.  In Fig. 23 is shown a form of sash bar without drip gutters, for use with butted glass.

As a rule, the lumber working factories do not have machinery for working (sticking) the drip grooves, and it will be necessary to obtain them from some firm dealing in green-house material. There are several large concerns who deal in cypress, and furnish everything required in the construction, including gutters, ridge, plates, rafters, sash bars, ventilating sash, doors, etc., all cut ready to put together.

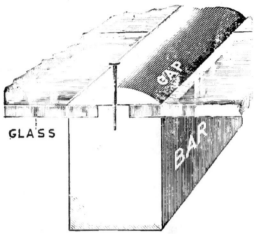

FIG. 23.    PLAIN SASH BAR FOR BUTTED GLASS (*Section*).

This will be a great help to the small florist, as he can secure his lumber of standard shapes and sizes, with plans that will enable any carpenter to put it together.

## PORTABLE ROOF.

An old plan of construction is to make a framework for the roof, with two by six inch rafters and a heavy ridge board (Fig. 24).  The roof is covered with movable sash, similar to hotbed sash, from three by six to four by eight feet in size.  If the house is narrow, one sash on each side will cover it.  The sashes may be screwed to the plate and ridge, and thus make a tight roof.  To secure ventilation, some of the sash may be hinged, either at the top, bottom, or sides, or they may be provided with stops that will hold one end in place, while the other is raised (see hotbed Fig. 85).

In wider houses, in which the rafters measure more than eight feet in length, the space between the top of the sash and the ridge may be covered with a smaller sash, the lower edge of which laps down upon the large sash beneath. Where two rows of sash are used in this way, it is customary to have all of the upper row hung on hinges (Fig. 24), although if they are very large, not more than every third one will be required for ventilating purposes, and the others can be screwed down. One

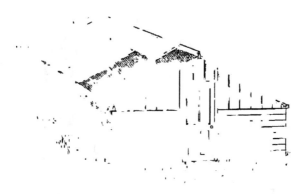

FIG. 24.    GREENHOUSE WITH PORTABLE ROOF.

great objection to this kind of a house is that the rafters obstruct the light and heat and as the glass used for the glazing of the sash is generally quite small, the sash bars and the sash frame will also be a serious impediment. Where only one row of sash on a side is required, this trouble can in a measure, be avoided, by dispensing with the rafter and carrying the sash to the ridge.

The foregoing remarks are applicable when the houses are of a temperate character, and, to a certain extent, for houses in which crops are forced during a part of the winter only, as in growing hybrid perpetual roses. As a rule, however, this style of house is not only more expensive to build, but, for the reasons given, it is less desirable than houses built with

## PERMANENT SASH BARS.

While many houses are built without rafters, the sash bars being all of one size, the usual forcing house construction is to have every fifth sash bar of the nature of a rafter, either two by four inches, or, in large houses, two by five inches. The ventilators are then placed in a continuous row on one, or both sides of the ridge, occupying a space from fifteen to thirty inches in width, each sash extending from one rafter to the next. When this construction is used, a two by four-inch header is mortised into the rafters just under the lower edge of the ventilator, and the sash bars are fitted into this, at their upper end, the lower end being nailed to the wall plate.

Another method of arranging the sash bars with a continuous line of ventilators, is to have all of the sash-bars run from ridge to plate, thus dispensing with the heavy light-obstructing rafters, with short headers between the sash bars, instead of the long ones between the rafters. These short headers should be grooved to receive the glass on the lower side. The bars in the ventilators should be arranged directly over the sash bars, but even then, this method of construction is often objected to, as obstructing too much light at the ridge. This fault can, in a measure, be overcome by cutting off every other sash bar, and supporting the headers between those that remain.

A modified form of the rafter construction restricts the ventilators to half the length of the ridge, and admits of sash bars running from ridge to plate in the remaining sections. One of the simplest methods of construction is to cover the entire roof with sash bars,

and then cutting off every eighth sash bar four feet
from the ridge, and inserting a grooved header to support it. This will provide for a ventilating sash two to
three feet wide, by four feet long, every eight or ten
feet, according to the size of glass used.

## RIDGE.

The ridge should be of either one and one-half or
two-inch stuff, and from six to eight inches deep, according to the size of the house and of the sash bars.
It should have a groove for the glass on one side, in case
there is but one line of ventilators, or on both sides if
the ventilators are not continuous. The arrangement of
the ridge is shown in Fig. 25. The ridge may be surmounted by a cap, and, particularly if the building is a
conservatory, an ornamental cresting, with finials at the
extremities, should be added. Even in case of commercial houses, their attractiveness is so much increased by
the addition of some simple forms of scroll finials, as
shown in Fig. 5, that the expense should not be considered extravagant.

## DETAILS FOR ROOF.

From Fig. 25 the details for the construction of an
even span house eighteen feet wide can be obtained, and,
with slight modification, they can be used for any other
form. In addition to an end view of the house, the following sections are shown: A side wall with gutter;
wall with side plate; ridge and ventilator; purlin;
double gutter for use when two houses are built, with a
wall in common; roof sash bar, and of end wall showing
gable rafter, end sash bar, and gable sill. The scale for
the elevation is three-sixteenths of an inch to the foot,
and for the details one-sixteenth of an inch to the inch.

In constructing the roof, the sash bars and end
rafters should be cut at such an angle as will make a

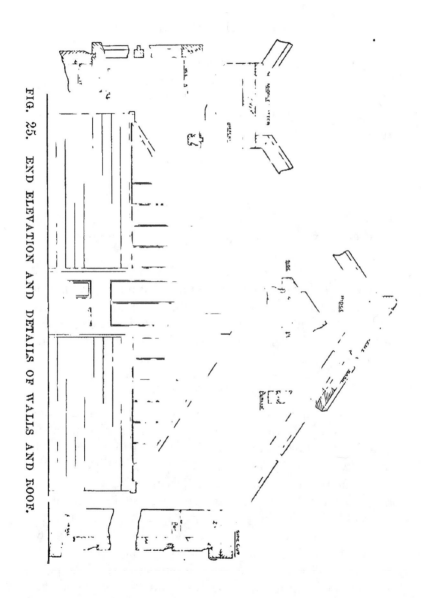

FIG. 25. END ELEVATION AND DETAILS OF WALLS AND ROOF.

tight joint with the ridge above and the plate below,
and then firmly nailed in place. If the plates are placed
at the same angle as the roof, the lower ends of the sash
bars should be let in to them about half an inch. As
the panes of glass are generally of scant width, if the
sash bars are spaced so that they are exactly as many
inches apart, measuring from shoulder to shoulder, as
the glass is supposed to be wide, a good fit will be
obtained.


## CHAPTER VII.

### COMBINED WOOD AND IRON CONSTRUCTION.

The use of iron for posts and rafters has been re-
ferred to, and, as the growing opinion among greenhouse
men is, that the question of durability should be considered more than it has been in the past, there can be no
question but that in the construction of greenhouses, in
the future iron will be quite largely used.

### IRON RAFTERS AND PURLINS.

Various methods of construction are now in use, one
of the best combining a framework of iron with wooden
sash bars. For forcing houses, the rafters are about
three by one-half inch, as shown at (1) in Fig. 26, and
are surmounted by a wooden rafter cap. The rafters (2)
are fastened to each other and to the ridge by iron knees
or brackets (3). The purlins are of one and one-half to
two-inch angle iron, and are fastened to the rafters by
means of iron lugs (4). If desired, gas-pipe purlins can
be used. With large glass, and small sash bars, the purlins should be quite near together, but as the size of the
sash bars increases, or that of the glass decreases, they may

be farther apart. While four feet will be none too little, in one case, they may be as much as eight feet in the other. When the ventilators are in long rows, either side of the ridge, the upper line of purlins should be under the lower edge of the sash, and should carry a wooden header, into which the upper ends of the sash bars are mortised. To the other purlins the sash bars are fastened by means of wood screws.

When the distance between the rafters or other supports is not over six or seven feet, one-inch gas pipes

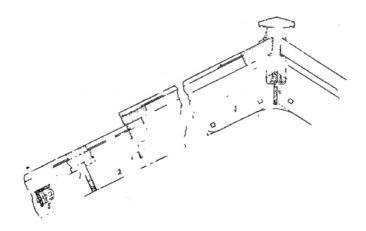

FIG. 26.   DETAILS FOR COMBINED IRON AND WOOD
ROOF.

will make quite a stiff roof. They can be inserted in holes in wooden rafters when these are used, or can be held up by means of small castings attached to iron rafters. When the roof is constructed of sash bars, without the use of rafters, a continuous line of pipe supported by posts, at intervals of six feet, will form a good purlin. Fig. 27 *A* shows a gas-pipe purlin, and *B* shows the clips for attaching the pipes to the sash bars. The pipe may be cut in lengths of six feet, and screwed into the tees to which the posts are attached, or, what is perhaps easier to put up, the tees are reamed out, so as to allow the pipe to slip through them. The lengths are screwed together, and, if desired, can be used as water pipes. If the purlin is connected by screw-joints with one or more of the posts on each side, a hose can be attached,

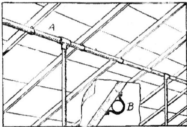

and, although the effect will not be lasting, the water contained in the pipes will have the chill taken off.

When a pipe purlin is used, with supports more than eight feet

FIG. 27.   GAS PIPE PURLIN. apart, it does not give good satisfaction, as it is more or less likely to sag. In order to hold the sash bars firmly down on the purlins, iron clips can be used, which should be screwed to, at least, every other sash bar.

### CENTER POSTS AND BRACES.

In narrow houses with a walk in the center, no center post need be used, as, if the wall posts are firmly set, and particularly if a truss bracket is used in the angle of the roof (Fig. 28), there will be no danger of its sagging. As the width of the house increases, a necessity arises for either supporting posts or truss rods.

FIG. 28.   IRON BRACKET FOR ROOF.

In wooden houses over fifteen feet wide, where there is no center walk, it is necessary to have a row of gas-pipe posts, either one inch, or one and one-fourth inches in diameter, to support the ridge pole, and if rafters are more than eight feet long, another row should be used to support them in the center.

In wide houses the rows of supporting posts should be about six feet apart, one for each purlin. When the posts would stand in the walk, if placed vertically, they may be arranged as braces from the center posts, either as shown in Fig. 29 or in Fig. 60. If the ridge is supported there will be no danger of the walls spreading, even if diagonal braces are used.

In one or two houses of recent construction the posts have been used as legs for the center bed, by inserting tees, into which the cross bearers for the bed are

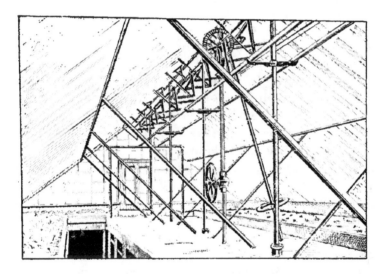

FIG. 29. IRON POSTS AND BRACES.

screwed. The upper ends of these posts are fastened, by means of top castings, to wood or iron rafters, or by means of the tees previously mentioned to the pipe purlins. The lower end of the posts may be inserted into cedar blocks, or rest on masonry piers, either upon flat castings (Fig. 14), or in beds of cement.

When iron rafters are used, particularly if there is a solid shoulder at the eaves, or if the roof is strengthened at that point by a strong angle bracket, there will be no

necessity for supporting posts unless the house is very
wide; but a truss rod, if necessary, may be used to keep
the roof from crowding the walls out.


## CHAPTER VIII.

### IRON HOUSES.

We have, thus far, only considered houses constructed of wood, or partly of wood and iron, but, for
many years houses built entirely of iron and glass have
been used in Europe, and they are now frequently seen
in this country. In favor of these houses it is claimed
that they are almost indestructible, and that, if the iron
is galvanized, there will be no necessity of painting the
houses. In some cases, zinc or copper is used for the
sash bars, and the same claims are made for those houses.
For the most part, these claims are true, and, although
one could afford to pay an increased price for iron houses
that would need no outlay for repairs or renewal, provided everything else is equally desirable, there are several serious objections to iron houses, that have, for the
most part, restricted their use to large conservatories,
and, even there, the combined wood and iron construction is fairly holding its own.

The objections may be stated as follows: 1st. As
iron is a rapid conductor of heat, the amount thus taken
from the house by the iron sash bars will be, perhap
three to five times as great as would be the case were
wooden sash bars of the same size used, and this requires
a noticeable increase in the amount of fuel consumed.
Several builders of iron houses, however, have so reduced
the amount of iron exposed to the outer air, that, so far
as radiation is concerned, there is, perhaps, no great
difference.

2d. With several of the methods of glazing, the packing used, although tight at first, soon becomes loose, and allows the heated air to escape through the cracks.

3d. Even if the roof is water-tight, there will be a large amount of water congealed on the under side of the sash bars at night, which, melting as the heat rises in the morning, causes quite a shower. Frequently, in systems where large glass is used, a metallic strip is placed between the panes to act as a gutter, to catch the moisture condensed on the glass. If it works all right there should be no drip from the glass, but they frequently become clogged.

4th. Even if such is not the case in England, it is found, in our extremes of temperature, that unequal expansion and contraction sometimes cracks the large panes, unless everything is very carefully adjusted, so that there is more or less broken glass.

These objections have most force with the sash bars used for skylight glass, in conservatories, and do not hold true to the same extent when used with smaller panes in forcing houses. In conservatories, however, although the drip is not desirable, it does far less injury than in houses used for forcing and growing plants, and one will need to place the greater durability and cheapness of maintenance of the metal roofs against the acknowledged increase of fuel required to heat the houses. The use of iron sash bars with metallic glazing, for commercial forcing houses, has not become general, as the matter of drip and of fuel, to say nothing of the increased first cost of the houses, are questions of considerable moment with florists.

## METALLIC SASH BARS.

Of the various forms of sash bars and methods of metallic glazing, the two that have been longest and most extensively used are the Helliwell patent system, controlled by the Plenty Horticultural and Skylight Works, and those in the hands of the A. E. Rendle Co.

### HELLIWELL PUTTYLESS SYSTEM.

The Helliwell system makes use either of a steel sash bar, as shown in Fig. 30, or of a zinc or copper bar, as in Fig. 31. The glass is held in place by long

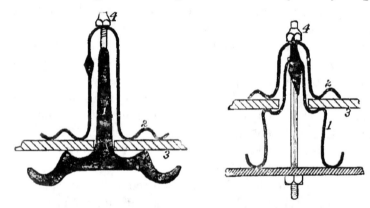

FIG. 30.   STEEL BAR.        FIG. 31.   ZINC BAR.
HELLIWELL PATENT GLAZING.

clips of zinc or copper, drawn down upon the glass by small bolts. It is claimed, by some, that the zinc bars are not stiff enough. The steel bar does not have this objection, but it is considerably more expensive. In Fig. 31 the sash bar is shown, resting upon a purlin, to which it is attached by a bolt. Candle wicking is used instead of putty.

### PARADIGM PATENT GLAZING.

The Paradigm system of glazing, used by A. E. Rendle Co., differs principally in the form of the sash bar and in the fact that the glass is butted. The sash bar is shown at A, Fig. 32. It is fastened to the purlins by lugs, as shown in the section. The glass rests upon the vertical sides of the sash bar, and is held in place by a copper cap, *D*, which is drawn down upon the glass by a small bolt, *C*. The sash bar serves as a gutter, to carry down to the plate any water that may enter between the cap and the glass. When sheet glass is used, all that is necessary is to put it in place and bolt down the cap. When large, rough plate glass is used, cross gutters are inserted between the panes. Directly beneath the points where the panes meet a section is cut out of the sash bars, and a piece of copper, bent as

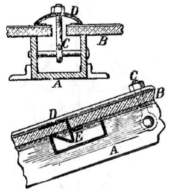

at *E*, is inserted. This not only catches any water that enters between the panes, but the condensed moisture on the inside of the panes is trapped, as it runs down the glass, and is carried to the gutter in the sash bars. If desired, white lead can be used in the joints, and an air-

FIG. 32. PARADIGM GLAZING. tight roof secured. There should be no sag in this sash bar, and it seems to have several features that are valuable.

## GALVANIZED IRON SASH BARS

Within the last few years, galvanized iron has come into use for greenhouse roofs. The framework consists of angle and T iron, put up in about the same way as when wooden sash bars are used. The ridge cap, cornice, gutters, and all exposed parts of the roof, are of galvanized iron.

One of the simplest forms of iron sash bars is shown in Fig. 33. It is made and used in the erection of conservatories, by M. H. Crittenden & Son, of Minneapolis, Minn. As will be seen, it much resembles, in

shape, some of the forms of cedar sash bars, and consists
of heavy galvanized iron, bent as shown in the illustration. At the lower edge are broad drip gutters, which
will not be likely to become clogged. The glass may be
laid, in any way desired, with putty. A V-shaped cap
(2) rests upon the top of the sash bar, and is held firmly
down upon the glass (3) by means of copper clips (4).
Unless the purlins are placed quite close together, it
would seem likely that the sash bars would sag, although

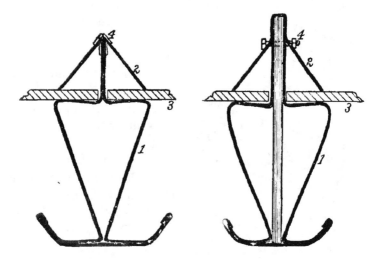

FIG. 33.   WITHOUT CORE.   FIG. 34. WITH STEEL CORE
GALVANIZED IRON SASH BARS.

a number of large houses are put up in this way, and are
said to be giving good satisfaction.

A form of sash bar that differs from the above, by
having a core of steel three-sixteenths of an inch thick
in the center to add to its strength (Fig. 34), is also
used. The clip holding the glass is drawn down by a
bolt. This, of course, is stronger than the other, but
the cost is more. From the form and method of putting
up these sash bars, there can be but little heat lost, from

radiation by the iron, and as the gutters seem to be arranged to catch all of the moisture condensed, there seems to be fewer objections to these sash bars than to almost any of the metallic sash bars.

## CHAPTER IX.

### THE PITCH OF THE ROOF

All plants require light, in order to assimilate their food; an optimum temperature is also desirable for the proper performance by the organs of the plants, of their functions. From the sun we obtain not only light and heat, but chemical or actinic rays, whose effect on plant growth is not well understood. In the case of greenhouse plants, the intensity of the sun's rays is greatly modified by the angle at which they strike the glass, as well as by the thickness and character of the glass itself. It has been found that about twelve per cent of the light rays are intercepted, in passing through ordinary sheet glass, and sixty per cent. in their transmission through opal glass.

This shows that much can be done by using clear glass to prevent the interception of the rays, and as the additional amount that is lost by reflection depends upon the angle at which the rays strike the glass, the careful adjustment of the slope of the roof should not be neglected.

#### REFLECTION AND REFRACTION BY GLASS.

When rays of light fall upon sheet glass at a right angle, they pass through without being turned from their course, and there is no loss, except from absorption, which will amount to about twelve per cent.

4

When they meet the glass at an oblique angle, a portion
of the rays are reflected, and the remainder, less those
lost by absorption, pass through the glass, and leave it
in the same direction they had before entering.

Fig. 35 illustrates the effect of a pane of glass, *x y*,
upon rays of light falling upon it at various angles, *A*
having ninety, *B* forty-five, and *C* fifteen degrees. *A*
passes directly through and emerges with eighty-eight
per cent. of its original intensity. *B*, on meeting the
glass, has four and one-half per cent. of its rays reflected
to *B'*; the balance, on entering the glass, are refracted,
or bent from their course, and, on leaving the glass,

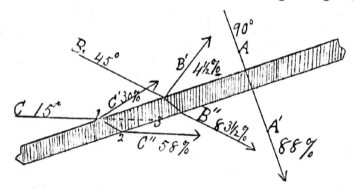

FIG. 35.    EFFECT OF GLASS AT DIFFERENT ANGLES.

with eighty-three and one-half per cent. of their first
intensity, are refracted, or bent back to their original
direction at *B''*. The effect upon the rays at *C*, which
meet the glass at an angle of fifteen degrees, is not
unlike that upon *B*, except that thirty per cent. of the
rays are reflected at *C'*, while only fifty-eight per cent.
emerge at *C'''*. The refraction, if anything, especially in
the case of very oblique rays, is a benefit. The absorption increases with the thickness of the glass, and it is
evident that there would be more loss were it obliged
to take the course 1-3 than there is in its refracted
course 1-2.

The following table gives the amount of light lost by reflection at different angles of incidence :

| Angle of ray 60 degrees. | Light lost 2.7 per cent. |
|---|---|
| "    "    50    " | "    "    3.4    " |
| "    "    40    " | "    "    5.7    " |
| "    "    30    " | "    "    11.2    " |
| "    "    20    " | "    "    22.2    " |
| "    "    15    " | "    "    30.4    " |
| "    "    10    " | "    "    41.2    " |
| "    "    5    " | "    "    54.3    " |

During the short days of winter, when the sun is only above the horizon for less than ten hours, as many of the rays should be trapped as possible, especially previous to ten o'clock in the forenoon, and after two o'clock in the afternoon. At the winter solstice, when the sun is farthest to the south, it rises about twenty-five degrees above the horizon at noon, and the slope of the roof should be such that the amount of light reflected while the sun is between the horizon and the above altitude, should be the least possible.

When the pitch of the roof brings the glass at an angle of twenty degrees, the sun, at five degrees above the horizon, will strike it at an angle of twenty-five degrees, and about sixteen per cent. of its rays will be reflected, in addition to, at least, twelve per cent. of the remainder, which will be absorbed in passing through the glass. Had the roof been given a pitch of thirty-five degrees, the sun at five degrees above the horizon would strike the roof at an angle of forty degrees, when only five and seven-tenths per cent. of the rays would be reflected, or only about one-third as many as were lost by reflection when the roof had a slope of twenty degrees.

## THE OPTIMUM PITCH.

It is evident, from this comparison, that there should be a slope of, at least, thirty to thirty-five degrees, to the roof, and that still better results in trapping the rays of light will be obtained if the roof has a slope to the south of sixty degrees, or more. The heat and actinic rays, in their passage through the glass, are subject to much the same laws of reflection and absorption as those of light; and in the case of absorption, the effect produced by semi-opaque glass is even greater. In determining the proper pitch for the roof of a greenhouse, in addition to considering the requirements for the transmission of the sun's rays in their full intensity, at the season when they are most needed, various practical considerations should be taken into account, among which would be the height of the side walls, the width of the house, the height of the roof above the plants, and the effect upon the heating of the houses, as well as upon the drip from the glass.

It will at once be seen that it is not desirable to have a roof so steep as to greatly increase the glass area, and, consequently, enlarge the consumption of fuel; while, if it is understood that plants grow best when comparatively near the glass, it will be seen to be unwise, except in "short span to the south" houses, to have the roof at a very sharp incline, as it will bring the plants in the center of the house at a considerable distance below the glass. With flat roofs not only is the rain likely to beat in between the laps of glass, but the amount of drip, from moisture condensed on the under-side of the panes, will be greatly increased. When a roof has a slope of thirty degrees (seven inches in a foot), or more, there will be no trouble, but at anything under twenty-six degrees (six inches in a foot) there will be more or less drip, both from outside and inside moisture.

The use to which the houses are to be put should also be taken into account, as if to be used only for wintering over plants, no growth being desired, it will be economy, both in construction and heating, to have the roof as flat as possible, and as good results will be obtained from a pitch of twenty-six degrees, as in a greater one. On the other hand, for crops that require an abundance of light for their quick development, the slope should not be less than thirty degrees, and if it can be secured without interfering in any way with the usefulness of the house in other respects, thirty-five degrees would be better.

## MEASURING THE PITCH.

The following table is given to show the angle that will be made by the sash bars for various widths of houses, and for different heights of ridge. In using the table, it must be understood that the width is measured from the bottom of the sash bar to a point directly under the ridge, while the height is measured on a plumb line from the upper end of the rafter to the level of the lower end.

ANGLE OF ROOF FOR DIFFERENT HEIGHTS AND WIDTHS.

| Width Feet. | 4 | | 5 | | 6 | | 7 | | 8 | | 9 | |
|---|---|---|---|---|---|---|---|---|---|---|---|---|
| | o | ' | | ' | | ' | | ' | | ' | | ' |
| 6 | 33 | 21 | 31 | 48 | 45 | | 49 | 24 | 53 | 8 | 56 | 18 |
| 7 | 29 | 44 | 35 | 52 | 40 | 46 | 45 | | 48 | 49 | 52 | 07 |
| 8 | 26 | 33 | 32 | | 36 | 52 | 41 | 11 | 45 | | 48 | 22 |
| 9 | 23 | 57 | 29 | 3 | 35 | 5 | 37 | 52 | 41 | 38 | 45 | |
| 10 | 21 | 48 | 26 | 33 | 30 | 8 | 35 | | 38 | 39 | 41 | 59 |
| 11 | | | 24 | 26 | 28 | 36 | 32 | 28 | 36 | 2 | 39 | 17 |
| 12 | | | 22 | 57 | 26 | 33 | 30 | 15 | 33 | 41 | 36 | 52 |
| 13 | | | 21 | 2 | 24 | 47 | 28 | 18 | 31 | 36 | 34 | 42 |
| 14 | | | | | 23 | 12 | 26 | 34 | 29 | 44 | 32 | 44 |

From the table it will be seen that in an even span house twenty feet wide (ten feet from plate to a point plumb with the ridge), a slope of about thirty degrees (30° 58') can be obtained by raising the ridge six feet above the level of the plate (the distances for both height and width being measured from the ends of the sash bars), while, if it is placed at a height of seven feet, a slope of thirty-five degrees will be obtained. In the same way, taking the figures for the width from the vertical column at the left, and the height for the ridge above the plate from the upper horizontal row, the num-

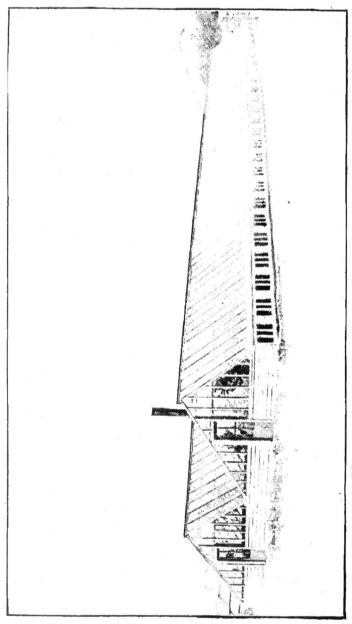

FIG. 36.   SHORT SPAN TO THE SOUTH CARNATION HOUSES.

*(Fred. Dorner and Son, Lafayette, Ind.)*

ber of degrees in the slope of the roof will be found where the corresponding lines intersect.

## SHORT SPAN TO THE SOUTH.

The above remarks apply, for the most part, to the pitch of the roof in even span, or in three-quarter span houses, when the long slope of the roof is upon the south side. It was stated, however, that if a slope to the south of sixty degrees could be obtained, more of the light and heat of the sun could be trapped. During the past two years several houses have been erected with a short span to the south and the long one to the north (Fig. 36), differing from three-quarter span houses turned half around only in having both walls of the same height. As will be seen from the engraving (Fig 60), the houses are built with three walks and two wide beds, the north one being slightly lower than the other. It can be seen at a glance that the plants upon the south bench are in an extremely favorable location, and can hardly fail to do well. The plants upon the north bed, however, are from eight to thirteen feet from the glass through which the sun's rays come, and are more or less shaded by the plants in the south bed. In theory, therefore, as a forcing house this form seems desirable, so far as the south bench is concerned; but for the north bench it does not seem, in any way, preferable to the even span house, except that the snow does not remain upon the steep south slope, so that there is less obstruction of light during the winter. In practice, however (which should be the real test), excellent results are claimed by Mr. George W. Miller, of Hinsdale, Ill., and by others who have tried it. As a summer greenhouse it has long been known that this form is a desirable one.

# CHAPTER X.

## GLASS AND GLAZING.

In no portion of a greenhouse have as great changes been made, perhaps, as in the glass and the method of setting it. A comparatively few years ago, glass as small as five by seven, and six by eight inches was used; it was usually of only single strength, and was of such poor quality that the leaves of the plants were badly burned. The panes were often lapped for an inch or more, and the putty was placed over, rather than under the glass.

The glass most commonly used to-day is known as sheet glass, either single or double strength. The latter costs somewhat more than the single strength, but it is less likely to burn the plants, and as it will stand a much harder blow, the breakage from hail storms and by accident will be much less, so that it will be cheaper in the end. In selecting greenhouse glass, two points should be borne in mind: (1) it should be even in thickness, flat, and free from imperfections that would cause sun burning; (2) the glass should be of good size.

### DIFFERENT GRADES OF GLASS.

Glass is graded as "Firsts," "Seconds," "Thirds," etc., the quality growing poorer as the numbers enlarge. The imperfections in glass are caused by air bubbles, unmelted specks, or various impurities. As the glass is melted, the impurities settle to the bottom, leaving the glass at the top quite clear. From this the "Firsts" or "Bests" are made; the "Seconds" come from a layer just beneath, and so on to "Fifths" and "Sixths,"

56

which are of quite poor quality. The lower grades are made by less experienced workmen than the "Firsts," and not only are they more likely to contain imperfections, but they are less even in thickness.

In the past, "Seconds" of French or Belgian sheet glass have been commonly used, and are still preferred by most builders, but American natural gas glass is now being extensively used, and it can be said that the "Firsts" are fully as good as French "Seconds," while the American "Seconds" make a very satisfactory roof. The grade known as "A" quality American glass is suitable for almost any purpose, while "B" quality will answer for many classes of houses. The natural gas glass is thought, by some, to be fully equal to the same grades of European glass.

## THE SIZE OF GLASS TO USE.

The size of glass has been on the increase, until to-day we find panes twenty, and even twenty-four, inches wide in use. While this extremely large glass makes a very light house, well suited for growing roses and lettuce, it is generally thought that a smaller size is preferable. For widths above eighteen inches the price rapidly increases, and this extra cost will be an important question, both at the time of erection, and in case of breakage. When the glass is to be butted, square panes are preferable, as it is likely to have straight edges at least one way. In sections of the country where the snowfall is heavy, the danger of loss from breakage increases as the panes are enlarged, and although twenty inch glass may be used in the South, eighteen inches will be a maximum width in the northern states, even for forcing houses, while, for ordinary florists' houses, the sixteen, and even fourteen, inch glass is regarded as the best to use, everything being considered.

Unless there is a decided change, the above widths, in lengths of from twenty to twenty-four inches, are the ones most likely to be used. This applies, of course, only to sheet glass, as rough plate or skylight glass and fluted glass may be used of a much larger size.

## FLUTED AND ROUGH PLATE GLASS.

The fluted glass has, perhaps, a dozen ribs to the inch, and is used, to some extent, for large conservatories. For houses of this kind, built with metal sash bars, it is, perhaps, preferable to either sheet glass or rough plate. The rough plate or skylight glass, as used in greenhouses, varies from one-eighth to one-half an inch in thickness, and from twenty-four by thirty-two to perhaps thirty-two by forty-eight inches. While well adapted for palm, and even for stove houses, it is not desirable for growing houses of any kind, as these, during the winter, need all the light they can wring from the sun.

The amount of light and heat absorbed by glass varies with its thickness, as well as its clearness, and as the fluted and skylight glass are both semi-opaque and quite thick, they will probably absorb fully half of the light and heat that enters them, to say nothing of what is reflected, and their thickness, although of advantage in giving them strength, is an objection in growing and forcing houses.

## DOUBLE AND SINGLE STRENGTH.

On account of the increased obstruction to the heat and light rays by the double strength sheet glass, as compared with thin panes, many prefer the latter for rose forcing houses, but it would seem that the amount lost by the necessity of bringing the sash bars closer together would more than counterbalance it.

While double strength glass costs somewhat more than the single, the greatly reduced loss in case of hail storms, and the fact that the breakage by frost and other

causes is less with the former than the latter, make it
preferable.  It is generally believed that, when in good
condition, the danger from hailstorms is only from one-
third to one-half as great.  The reports of the Florists'
Hail Insurance Association show that, although the
amount of double strength glass insured is in excess of
the single thick, the amount of glass broken is never
more than two-fifths as much, and in some years the
ratio is one to one hundred in favor of double glass.


# CHAPTER XI.

### GLAZING—METHODS AND MATERIALS.

In setting the glass, the end desired is to so arrange
it as to have the roof as nearly air and water-tight as
possible, and to have the glass held firmly in place.  As
usually laid, the glass is lapped, with the upper pane
extending about an eighth of an inch over the one below
it.  For curvilinear roofs this is practically a necessity,
and when the glass is straight and even, and well laid,
it makes a good roof.  Nearly all panes are more or less
curved, and if two panes in which the curves are not
equal are placed together, there is likely to be a crack
either at the corner or in the center of the panes.  Care
should therefore be taken to assort the glass and, if the
curves are of different angles, it is well to select those of
one angle for one row, and the others for another.

### PUTTY.

For glazing on wooden sash bars, if the glass is to
be lapped, astrals should be selected with half inch rabbets (Fig. 18), which should first receive a line of putty
sufficient to fill the shoulder.  The best grade of **putty**

should be used, and this should be mixed with pure white lead, at the rate of one part of lead to five of putty. If a larger proportion of lead is used, it will make the task of cleaning the bars a difficult one, in case of breakage, while, if the bars are kept properly painted, the mixture, as above, will hold for many years.

The putty should be worked rather soft, using linseed oil if necessary, and it will be found to stick to the wood best if it is as soft as can be used without sticking to the hands when they are well coated with whiting. Having applied the putty to a number of sash bars, the glass is laid on and carefully pressed into place, squeezing out all surplus putty until the upper end of the pane rests on the bar, and the lower upon the pane below, with a lap not exceeding an eighth of an inch. Care should be taken to have the curve of the glass up, if drip gutters are used, and down if they are not. The surplus putty, both inside and out, is then scraped off, taking pains to fill any cracks that may be left. With the old method of placing the putty on the upper side of the glass, it was found that in one or two years the water worked under the putty and it scaled off, leaving a crack at the side of the pane, as well as underneath. This both allowed the heat to escape and the water to enter, besides permitting the glass to slip down or blow off, if its other fastenings became loosened.

### GLAZING POINTS AND BRADS.

For holding the glass in place there are a dozen or more kinds of points and brads. One of the best seems to be an ordinary five-eighths inch wire brad (Fig. 37 A). This is stiff enough to hold the glass firmly in place, and has such a hold upon the wood that, if properly driven in, there need be no fears of its loosening and allowing the panes to slip down. Another advantage of this brad is that it is inconspicuous, and, consequently, not unsightly, and it offers little obstruction to the brush when the sash bars are painted. One of the most commonly used glazing points is cut from thick sheet zinc, and appears as in Fig. 37 *B*. In shape they resemble three-fourths inch shoe nails, which are also sometimes used. When driven well in, this form of brad has a firm hold, and, moreover, is quite stiff; the blunt end, however, tears its way into the wood, and, unless driven home, is readily detached. It is also more conspicuous than the wire brad, and is a slight hindrance to the painting. Two of these brads are used to hold the lower corners of the glass down in place, and two others are placed about an eighth of an inch from the upper edge, where they serve to hold the pane in place and to keep the pane above from slipping down. Large panes require two other brads in the center.

Of the various points used for glazing, none is better than the zinc triangle, No. 000 (Fig. 37 *C*). While

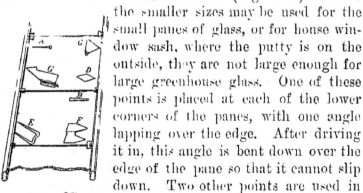

FIG. 37.
GLAZING POINTS.

the smaller sizes may be used for the small panes of glass, or for house window sash, where the putty is on the outside, they are not large enough for large greenhouse glass. One of these points is placed at each of the lower corners of the panes, with one angle lapping over the edge. After driving it in, this angle is bent down over the edge of the pane so that it cannot slip down. Two other points are used in the middle of the panes. The diamond points (Fig. 37 *D*) are driven in very rapidly with a machine, but are rather small for large panes, except when the glass is butted. Another point that is sometimes used is a double-pointed carpet tack. This holds the glass firmly in place, but it is not particularly ornamental.

Van Reyper's glazing point (Fig. 37 *E*) differs from the above in being bent in the center, so as to better fit the lower edge of the pane, and to this extent it seems to be an improvement. Eames' glazing point (Fig. 37 *F*) is double pointed, and is designed to both hold the panes down in place and to keep them from slipping, and it successfully accomplishes it. Ives' point (Fig. 37 *G*) has a single point, with one corner bent to prevent the slipping of the pane. It is rather thick, and as it tears the wood when driven in, it does not have a very firm hold, even with the shoulder at the point. One objection to the last two kinds of points is that they are "rights and lefts," which leads to more or less confusion in using them, and another which applies to all double-pointed points, is that in order to hold the pane securely they must be very accurately driven into place.

## BUTTED GLASS.

The method of setting greenhouse glass to which this term is applied, has been frequently advocated, and has been used, to some extent, for many years; but it has never come into general use, principally on account of its being somewhat more difficult to reset broken glass and make a good joint, than when the glass is lapped. This kind of glazing has many advantages over the other, among the more important of which are, that a tighter roof can be made, thus effecting a saving in fuel; there is less danger of broken glass, either from ice forming between the panes when lapped, or from accidents, as, when a lapped pane is broken it frequently cracks the one beneath; more benefit can be derived from the sun, as with lapped glass soot and dirt collect between the laps, causing an opaque streak, and even when this is not so, the double glass at the lap obstructs more light than the single glass. Moreover, admitting the fact that it is sometimes hard to get a good fit in repairing butted glass, using the old method of glazing, the labor of keeping a butted roof in good condition is less than for caring for one that is lapped, as there will be fewer breaks to repair, and using the new styles of sash bars the panes can be very readily replaced.

The only objection to butting the glass in glazing is, that upon flat roofs, after the glass has been set a few years, water, in a driving rain storm, will find its way between the panes and cause a good deal of drip. On the other hand, upon roofs with an angle of 35° or more, there will be sufficient adhesion between the water and the glass to cause it to run down on the under side of the panes to the plate, and thence to the ground, or, as arranged in some houses, either into an inside gutter, or through the wall into the outside gutter.

In laying glass upon the old style of sash bars, a thin layer of putty, or a film of thick paint, is placed on the sash bar, upon which the panes are laid and tacked in place, taking care to securely fasten the bottom pane in each row, to prevent slipping. In order to make the roof both air- and water-tight, it is well to seal the crack with white lead. To do this mix pure white lead with equal parts of good putty; spread this in a thin layer on a smooth board or pane of glass, and press the lower edge of glass against it before placing on the sash bars.

In setting the panes, crowd them together so as to force out all surplus material, leaving the lead to fill any inequalities between the panes and act as a cement to unite them. When this is properly done the rows of glass will virtually consist of a single pane, and will remain for several years, both air- and water-tight. In time the lead will work out of the larger cracks, but if they are so large as to prove troublesome they can be refilled with but little trouble. To make a good job in butting glass, all panes with rough edges should be rejected, or used only at top and bottom.

Having the panes nailed in place, the cracks at the sides should be filled by applying thick paint with a brush, or, as is preferred by some, by use of a putty bulb. The name of paint bulb would, perhaps, be as appropriate for the latest forms, which have a small brush projecting beyond the end of the tube, by which the crack is filled, and the surplus material brushed off (Fig. 38). If very much paint is used it will be necessary to sift on sand to keep it from running, but, when properly done, there will be little need of using sand upon it. If desired, the use of paint or putty under the glass can be dispensed with, although unless the glass fits snugly it will lessen the amount of paint that runs down between the panes. Ives' putty machine (Fig. 39) is very convenient for back-puttying in repairing roofs. Upon the more recent forms of sash bars the glass may be laid in paint or putty if desired, and the crack at the side filled in the same way; or both may be dispensed with, and the glazing performed by merely laying the panes in place on the sash bars (filling the cracks between the panes with white lead, if desired), and fastening the wooden strips in place by means of screws, thus holding them down.

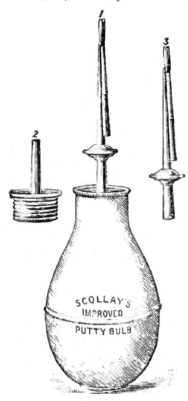

FIG. 38.    PAINT BULB.

## GLAZING STRIPS.

For use with this method of glazing, Gasser's glazing strip is considered very valuable by many who have tried it. It consists of a narrow strip of zinc bent into the form of the letter Z, as shown in Fig. 40, which is placed between the panes so that one leg of the Z is under the upper panes, and the other over the under ones. The cracks between the glass and the strip should be filled with white lead, or some other lasting cement, which will fasten them together,

FIG. 39. IVES' PUTTY MACHINE.

and thus make a tight joint. This will make a roof water-tight much longer than when the lead alone is used between the panes. If the strips are not properly laid, or if they are not cemented securely to the glass,

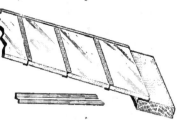

FIG. 40. GASSER'S GLAZING STRIP.

the leakage will be much greater than when no strips are used. Aside from their cost, and the labor of putting them in, the strips obstruct a small amount of light, but with large panes none of these objections are of serious importance.

From the present light that can be obtained on the subject, the best advice as to glazing of greenhouses and forcing houses is, use one of the sash bars shown in Figs. 20, 21 and 22; have the roof with an angle of thirty-five degrees; butt the glass, closing the crack with white lead, or, if a roof that will remain water-tight for many years is desired, use the glazing strip. With glass of a

5

width greater than sixteen eighteen inches, it will be best to lap the panes. When butted glass, laid with the convex side down, is used, there will be no necessity for drip grooves in the sash bars upon steep roofs, if there are no cracks at the sides of the panes. One important feature of this method of glazing is, that when resetting broken glass, instead of bothering to fit the panes, as is necessary with ordinary sash bars, one needs only to loosen the screws that hold the cap, and, slipping up (or down) the remaining panes, place the new one in place at the bottom (or top), and screw down the cap; or, if the panes are cemented in place, one can be selected that will fit the opening.

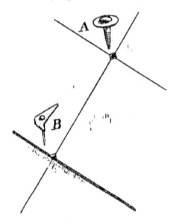

FIG. 41. NEW METHODS OF GLAZING.

To be used successfully, glass, to be butted, should be true and even; as, if panes with different curves are placed together, water will be collected and drip, unless the roof is quite steep. The difficulty increases with large panes, and sizes over sixteen inches will need to be very carefully selected, if used in this way, even with the glazing strip.

### NEW METHODS OF GLAZING.

Two other systems of glazing are shown in Fig. 41, one of which is for butted glass, and the other for lapped glass. In both, the sash bars are used without rabbets, which makes a lighter roof than can be obtained in any other way. In the first method, which was used by J. D. Raynolds, of Riverside, Ill., the glass is butted, as shown at A, and is held in place by a screw and washer, at the intersection of the panes. By breaking off a small corner from each, the screw can be inserted, and the washer will press the glass into place. By the other method, which was described in the *American Florist,* the glass is lapped, and held in place by a piece of sheet lead, bent as at *B.* The lower corner of the panes should be nipped off, and an opening made through which a brad or screw can be inserted. If desired, a film of white lead can be placed between the panes to close up the joints, but no other painting will be necessary upon the exterior.

## CHAPTER XII.

### VENTILATORS.

For all kinds of plants it is desirable, at some seasons of the year, that means be provided for supplying fresh air, and for removing surplus heat. It has been found that, if openings are provided for the egress of the air, fresh air can find its way in, and no necessity will exist for considering that side of the question, except during the summer months. As the air of greenhouses is generally warmer than that outside, it will naturally tend upward, and ventilation will be most effective if provided at the highest part of the building. The ventilators should be arranged so as to prevent direct drafts of cold air upon the plants. They are sometimes placed on both slopes of the roof, in order that the opening may be opposite to the direction of the wind.

In some houses large ventilators have been placed, at intervals, along the roof; but better results are obtained when continuous lines of narrow ventilators on one or both sides of the ridge are used.

### CONTINUOUS VENTILATION.

When a continuous row of ventilating sashes is used, a small opening will provide the necessary ventilation; but, if they are scattered at intervals along the roof, the openings will need to be two or three times as large, and the draft of cold air upon the plants will be greatly increased. The openings at the ends of the sash invite side drafts. It is a poor plan to have a continuous row

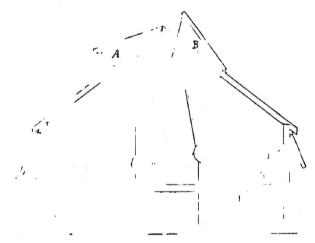

ARRANGEMENT OF VENTILATORS.

only part which are used. Particulars

shafting is necessary to work

Although necessary

joints should be located over the middle of rafters or sash bars. The glass used for the sash should be of the same width as for the rest of the house, except the rows at either end of each sash, which should be somewhat narrower, to allow for the increased width of the side strip of the ventilating sash.

## HANGING THE SASH.

The old method of hanging the sash was to have the hinges on the upper side (Fig. 42 A), but as, for the same size of opening, a ventilator will be more efficient when hinged at the lower edge (Fig. 42 B), that method will be generally used, especially when there is only one line of sash. When only one line is used, they should be on the same side of the roof as the prevailing cold winds come from, when hinged at the bottom, and on the other side if hinged at the top.

## VENTILATING MACHINERY.

In small houses, a simple method of opening the ventilators is by means of what are sometimes called skylight fixtures, which are fastened to the lower edge of the ventilator by screw eyes. They have holes at intervals, through which a pin on the edge of the header is passed, thus holding the sash at any angle desired. One sash at a time only can be opened, and, for houses of any length, some form of apparatus that will open all the ventilators on a given line is desirable.

## A SIMPLE APPARATUS.

One of the simplest is shown in Fig. 43. It consists of lifters made of one inch by one-fourth band iron (B), about two feet in length, fastened rigidly to the lower edge of the ventilator (A), and extending down into the house at right angles to it. A small wire cable runs the length of the house, and near each ventilator a cord (C) is attached, which, after running through a

pulley, is fastened to the lower end of the lifter. The
cable is arranged so that it can be readily drawn through
the house, lifting all of the sash to any required height.
The motive power may be applied to a small rope running through pulley blocks, or by means of a small
windlass. As first made, they were closed by their own
weight, and, as they were not held down in any way,
accidents often happened in high winds. An improvement (Fig. 43 *D*) is in an additional rope, attached to
the bottom of each sash, and running through a pulley
to a point beyond, where it is fastened to the main cable.
If the cable at the farther end of the house is carried

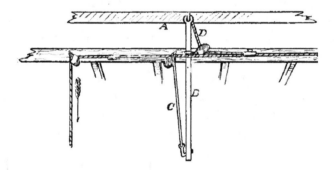

FIG. 43.    A SIMPLE VENTILATING APPARATUS.

over a pulley, and has a heavy weight attached to it, the
cable will be drawn back, when it is desired to close the
ventilators, and will hold them securely in place.

### SHAFTING.

In nearly all other ventilating machinery, the power
is conveyed by means of a gas pipe shaft running along
under the ventilators. In some cases it passes through
a cross placed in the ridge post, about a foot from its
upper end, and in others, it is held in place by means
of a clamp fastened to the post. The usual method of
fastening is by means of small hangers screwed to the

rafters. When sashbars alone are used to form the framework of the roof, some method of hanging the shaft to the posts is desirable, but not necessary.

Various methods of applying the power have been used, the most common of which is the common elbow

FIG. 44. NEW DEPARTURE VENTILATING APPARATUS.

fixture shown in Figs. 42 and 46. These are strong and do their work well, except that they are applied at a dis-advantage when the sash is just beginning to open, and as this is the point at which they are most frequently used, it will have to be regarded as a slight objection. Another form of fixture is regarded as a "New Departure," by J. D. Carmody, the inventor. It has much the shape of a meat saw (Fig. 44), and lifts the sash by the action of the cogs on the shafting, upon those on the fixture. It works easily, and the same force is required to lift the sash, at whatever height it may be. A form quite similar to this is used with the Little Victor machine, and is recommended by the inventor, E. P. Hippard, as valuable for small sash.

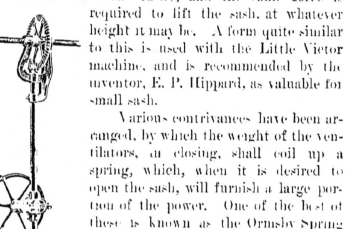

Various contrivances have been arranged, by which the weight of the ventilators, in closing, shall coil up a spring, which, when it is desired to open the sash, will furnish a large portion of the power. One of the best of these is known as the Ormsby Spring balance. It has proven quite satisfactory, and works very smoothly, opening, with the expenditure of very little power, a long line of sash. It is easy to put up, and the only objection to the system is that, after two or three years of use, the springs wear out.

Of the machines used to work the shafting, with its elbow fixtures, the simplest is the kind generally used by

FIG. 45.
STANDARD.

greenhouse builders. It consists of a large wheel upon the shaft, worked by a worm upon the upper end of a rod, to which the power is applied by means of a crank, or a hand wheel. It will be found in several of the illustrations. Among the more recent candidates for favor are the Standard and Challenge machines.

The former, shown in Fig. 45, is manufactured by E. Hippard, Youngstown, Ohio. It is a very easy work-

ing machine, but does not work quite as fast as some
of the others. In the old machines, the large wheel on
the shaft sometimes slipped, or was pushed away from
the small pinion, but with the new double header there

FIG. 46.   CHALLENGE.

should be no trouble. As will be seen from the engraving,
the power is applied, by means of a hand wheel and worm,
to the vertical shaft which works inside the post. At

the upper end of the shaft is a pinion, or bevel gear, by which the power is conveyed to the shaft. We have found it a very satisfactory machine.

The Challenge machine (Fig. 46) is made by the Quaker City Machine Co., of Richmond, Ind. It differs

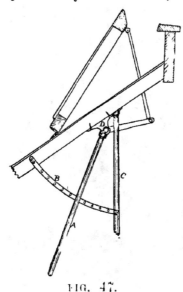

from the Standard in using sprocket wheels and an endless chain, instead of a vertical shaft and gearing. The first machines made had several slight defects, but these have been corrected in the latest pattern, and the machine now does very good work. While it does not work as easily as the Standard, it gives a more rapid movement to the sash, and to that extent is preferable to that machine; the latter excels in ease of operation, although either can be changed by varying the size of the gear and sprocket wheels.

FIG. 47.

A SIMPLE FIXTURE.

### CHEAP VENTILATING MACHINE FOR LOW HOUSES.

In low span roof houses a simple method of working the shaft is shown in Fig. 47. It consists of a lever of one inch gas pipe (*A*), perhaps four feet long, fastened to the shaft (*D*) by means of a T. By means of this, a line of narrow sash fifty feet long can be opened or closed, and can be held in place by means of a pin passed through a strip of iron (*B*), as shown in the sketch.

### OUTSIDE SHAFTING.

When the ridge is too low to admit of running the shafting under the roof, it may be placed on the outside, as shown in Fig. 48. In any house, if, for any reason, the shafting is not desired on the inside of the house, this arrangement, which is used by Hippard and others, may be employed.

While a single wide line of ventilators will answer, in houses less than twenty feet wide, two narrow lines on opposite sides of the roof will be preferable. In wide

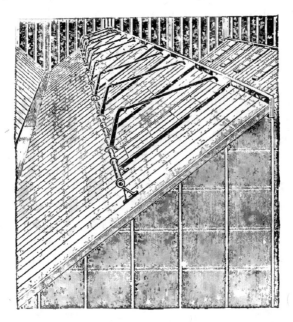

FIG. 48. OUTSIDE SHAFTING.

houses ventilating sashes in the vertical walls are desirable, and some builders of three-quarter span houses prefer to have one row at the ridge and one in the south wall, even in rose houses, to having the two rows at the ridge. In even span houses there is less occasion for side ventilation, except in wide houses. In Fig. 42 is shown a variety of methods of arranging the ventilating sash, and of attaching the ventilating machines.

# CHAPTER XIII.

## GREENHOUSE BENCHES

The benches used by the average florist are a constant source of trouble and expense. Built as they usually are, of cheap or waste lumber, their life is a short one, and they frequently break down while in use, either ruining the plants, or so mixing the varieties as to make them of little value. A wooden bench generally has to be renewed within five years, and in some cases three years sees their period of usefulness at an end. Not only is the florist required to pay for labor and material for constructing a new bench, but, as frequently the weakness is not discovered until the bed is being prepared for use, the delay necessitated in planting the bed may greatly lessen the profits. The best materials for greenhouse benches are iron and tile, or slate, but, as the average florist will think he cannot afford this kind of a table, let us consider the next best thing.

## WOODEN BENCHES.

With a little attention in constructing and caring for wooden benches, their durability can be doubled. When wooden legs are used, they should be raised above the level of the soil and walk upon a brick or stone pier. This will not only furnish a firm support, and thus prevent the settling of the bench, but it will serve to keep the lower portion dry, and check its decay. Red cedar or locust will make the most desirable legs but, as they cannot always be readily obtained, cypress or pine will be generally used. When the walls of the house are of

76

wood, the back ends of the cross bearers can be nailed to the posts, or to the studding, and in houses constructed upon a brick or grout wall this can readily be used for supporting them. Measuring from the inner face of the wall, the length of the cross bearer should be two inches more than the width of the bench, thus admitting of a free circulation of air in the rear. When the wall cannot be used to support the backs of the side benches, wooden legs can be used, the same as for the fronts, and for the middle bench. These should be about two by four inches, and from two to five and a half feet in height, according to location and the character of the house.

The cross bearers may range from two by four inches for narrow benches, to two by six, or even two by eight inches for wide ones. If fastened to the legs, as shown in Fig. 49 A, there will be little danger of their becoming detached and letting the bench down. If these supports are placed once in three and one-half or four feet, the common six by one inch fence boards can be placed longitudinally for bottoms, when shallow rose beds are desired, although the twelve-inch boards are better for the staging of pot plants. To provide for thorough drainage, cracks three-fourths of an inch wide should be left between the boards. If deep beds are to be used, or if large pot plants are to be placed on them, the distance between the supports may profitably be reduced to three feet, or one and one-half inch boards used. For the front and back of the benches strips of board from three to six inches wide, in accordance with the kind of

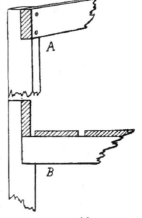

FIG. 49.

WOODEN BENCHES.

plants to be grown on them, should be used. If the
legs are extended to the top of the front boards, as in
Fig. 49 B, they will be held firmly in place and will last
much longer. The legs and cross bearers, as well as the

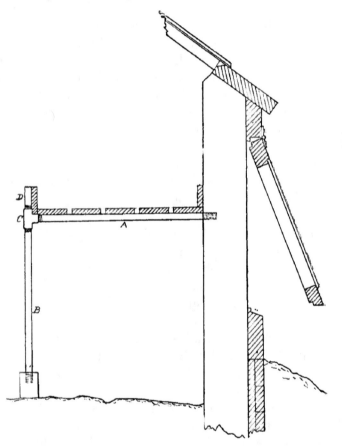

FIG. 50. GAS PIPE BENCH SUPPORTS.

side pieces, should be thoroughly painted before they are
fitted together, and this will often double their period of
use. After the bench is completed, it will pay to give
the inside a thorough coating with Louisville cement, of
the consistency of thick paint.

The width of greenhouse benches varies, to a large degree, with the width of the house, and the use to which it is put. The side benches, in a rose house, are sometimes as narrow as two feet and six, or, perhaps, eight inches, and are seldom wider than three feet and six inches, which should be the maximum. The center benches range from five to seven feet in width. When properly built and well cared for, benches of this description will be far more economical, in the end, than the cheap constructions generally seen in greenhouses.

### IRON BENCHES.

Many florists who are not ready to try the iron and slate bench, are using iron legs and cross bearers. The simplest forms are made of one-inch gas pipe (Fig. 50), the lower end of the leg resting in a cedar block sunk in the ground, and the upper end supporting the front end of the

FIG. 51. MENDENHALL'S BENCH.

cross bearer, by means of a malleable iron T, from the top of which a short piece of pipe extends even with the top of the front boards, thus holding it in place. The front boards can, if desired, be placed outside the pipe, and held in place by iron clips. The rear end of the cross bearer is screwed into the wall post, or set in the masonry, when possible, and if neither of these methods of support can be used, gas pipe legs can be provided, the same as for the front. For center benches a somewhat heavier construction would be necessary, the cross bearers being of one and one-fourth inch gas pipe. When houses are built on the "ridge and furrow" plan, the cross bearers for the side benches may pass through the wall into the adjoining house.

### ANGLE IRON BENCHES.

One of the simplest forms of iron benches was recently figured in the *American Florist* (Vol. VI, Page 983), as Mendenhall's bench (Fig. 51). It rested upon brick piers, and consisted of two or three inch angle irons, placed so as to form the front and the back of the bench. The bottom was of slate or boards, as desired. By using intermediate strips of T iron, narrow strips of slate, or bench tile could be used.

Another form of greenhouse bench has been tried in the houses of E. G. Hill, Richmond, Ind. The cross bearers and longitudinal strips are of a light street car T rails, with bottoms of slate and sides of narrow boards,

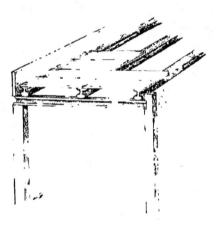

held in place by narrow strips of iron. The bench is supported on cedar posts (Fig. 52). The cost of the iron rails is given as eleven and one-half cents per foot. This is considerably more than the cost of the higher grades of angle iron, but as the rails are much stronger, the number of legs and cross bearers is reduced, which might easily make the iron bench the cheaper of the two. The slate and glass later to be considered is much in its favor.

Perhaps the cheapest form of iron bench is shown in Fig. 53. The size of the iron required is consid-

ing to the width of the bench, and the use to which it is
to be put.  For a rose house, the side benches can be
supported on one and one-half inch cross bearers of T
iron (*D*, Fig. 14), placed once in four feet, with one and

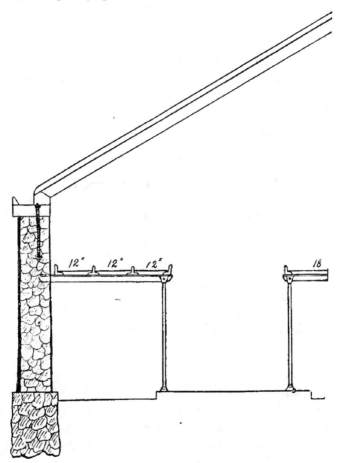

FIG. 53.   ANGLE IRON BENCH.

one-half inch angle iron (*G*) for the front and back, and
two intermediate one and one-fourth inch T irons (*H*).
Using twelve-inch slate, or tile, for the bottoms, the
bench will be three feet and two inches wide.  For the

6

legs, one inch gas pipe (*E*), or one and one-fourth inch T iron can be used. The gas pipe can be flattened at its upper end and bolted to the cross bearers, or it can be inserted into a casting (Fig. 14 *F* and *F*"), which can then be bolted on. With such a casting at the top, and a flat plate for the leg to rest in at the bottom, a very neat bench can be made. If sides are desired to the benches, larger angle iron can be used for the outer edges of the benches, say three by two inches, or three to five-inch strips of board can be used, and held in place either by the edges of the angle iron, or by means of screws put through holes in the angle iron.

### BENCH BOTTOMS.

The bottom is, as a rule, the first portion of the bench to decay, and if any part is to be of indestructible materials, this should be the one. The most satisfactory

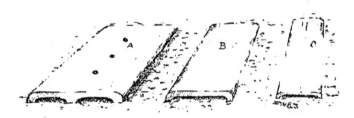

FIG. 54. TILES FOR BENCHES.

bench bottoms, in every way are some of the forms of "bench tile." They are more or less porous, and provide both for drainage and for a thorough aeration of the soil. Those invented by W. P. Wight, of Madison, N. J., seem particularly desirable. They can be made of any size desired, although about twelve by six inches seems a good one, and differ from most of the others in having a row of holes along the center (Fig. 54 *A*). The form shown in Fig. 54 *C* is five inches wide by twelve long, and Fig. 54 *B* represents a tile seven by

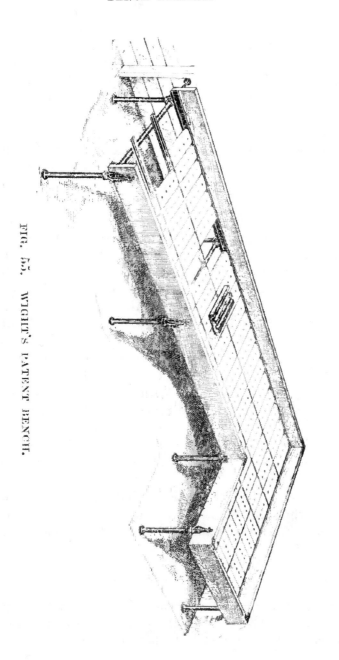

FIG. 55. WIGHT'S PATENT BENCH.

twelve inches, both of which are manufactured for "fire-proofing" the structural iron in modern fireproof blocks, but answer very well for bench bottoms. By leaving spaces between the tiles, ample drainage can be secured. Mr. Wight has invented a bench (Fig. 55) to be used with his tiles.

For the tables in the show house, or conservatory, upon which large plants only are placed, large slabs of slate, of the full width of the bench, may be used without any covering. In the growing houses some covering for the slate is desirable, and smaller sizes may be used. Heavy roofing slate, about twelve by eighteen inches in size, can be cheaply obtained, and, with a covering of coarse gravel, makes an excellent plant table. When used as bottoms for the tables in rose houses, and

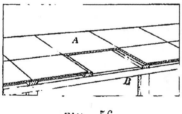

FIG. 56.

WOOD AND SLATE BENCH.

for cutting beds, they are less satisfactory than bench tile, as they allow of but imperfect drainage and aeration, and the soil and sand soon become sour. With careful watering the injury will be somewhat lessened, but the tile will be found more satisfactory. In Fig. 56 will be seen a method of using slate for bench bottoms, with wooden supports.

The use of boards for bench bottoms may be economical where lumber is cheap, and the other materials expensive, but the more durable materials will generally be preferable.

### SOLID BEDS.

For growing many crops the so-called solid bed will be desirable. These are of the same widths and in the same places as the raised beds, but, as a rule, are not as

high above the walks. When the underlying soil is
light and sandy, the application of about six inches of
prepared compost will be all that is required to make
them, except the erection of barriers of wood or brick to
keep the soil in place. It is generally necessary to provide some kind of drainage, and for this purpose three-
inch drain tiles have been found excellent. By placing
them two feet apart and eight inches below the surface,
across the beds, ample drainage will be provided, and the
warming and aeration of the soil will be promoted.

## CHAPTER XIV.

### PAINTING AND SHADING

In order to preserve the wood from decay, and the
iron work from rusting, the materials should be covered
with some substance that will render the woodwork
water proof, and prevent the oxidation of the iron.
There are on the market many patent paints that may
be suitable for certain purposes, but very few of them
will prove satisfactory for greenhouse painting. If pure
white lead and linseed oil, with a small amount of Japan,
are used, the results will be as satisfactory as can be
obtained from any mixed paint, and as the covering of
the framework of a greenhouse with some paint that
proved worthless for the purpose would lead to a large
expense, it is better to take something that is known to
be good, than to experiment with materials that, though
apparently cheap, may prove dear in the end. There
are several low priced brands of so-called "white lead,"
that are composed largely of zinc and baryta, and if
these are used the paint will peel off within a year.

If the houses are to be painted white, a little black should be added to take off the glare. However, some other light color may be preferred to white, and a pleasing one can be made by adding yellow and a small amount of green, producing a very light shade of green. With darker trimmings upon the house, this will be found quite satisfactory. While it is desirable to use pleasing tints for painting the greenhouses, preservation of the timber is the main object to be sought.

### PAINTING THE GREENHOUSE.

The priming coat should be given before the house is erected. As soon as the parts have come from the mill, the joints should be made, as far as is convenient, and, if possible, the woodwork should then be soaked in hot linseed oil. A long tank should be made, and by placing the oil in it the parts can readily be dipped. If steam pipes can be run through it, all the better. When this cannot be done, the woodwork should be given a thorough priming coat. The addition of yellow ochre, or some similar material, to the oil, will be of advantage.

In putting up the house, too much pains cannot be taken in coating every joint with pure white lead paint. The average carpenter will not see the advantage of this, but a coat of thick paint should be insisted upon.

As soon as the framework is up, a second coat should be given it. Our best greenhouse builders use two coats of paint for commercial, and three for private greenhouses. If only two coats of paint are to be given, every crack and nail hole should be filled with good putty before the second coat is applied, but, if a third coat is to be given, the puttying should be delayed until the second coat is dry. When three coats are to be given it will be best to apply the last coat to the inside of the house before the glass is on, although it would serve to hold the putty in place under the glass **if** it were applied after the glazing is completed. Whatever the number of coats, the last one to the exterior should not be given until the glass has been set, as then any crack that may remain at the sides of the panes can be filled, and the roof will be made water tight. The putty would also become softened, and would work out were it not painted.

In drawing the sash, on the exterior, the paint should be rather thicker than is used for ordinary painting, and it is an excellent idea if it is drawn out upon the glass for, perhaps, an eighth of an inch. In this way, the paint will serve as a cement to hold the panes in place, should the other fastenings become displaced.

## REPAINTING.

Whether two or three coats of paint are given the houses at the time of erection, another should be applied after one year, to the exterior, at any rate, although when three coats have been used the painting of the interior may be delayed another year. In order to keep a greenhouse in the best repair, one coat should be given to all exterior wood work each year, and to the interior every second year. This frequent application of paint is made necessary by the fact that, if cracks open at any place, water will enter, and the rapid decay of the woodwork will follow. If painting is long delayed, cracks large enough to admit water often open between the glass and the putty, and the latter, becoming softened, is washed out. Through the openings thus formed, heat will escape, and water can gain entrance.

### PAINTING IRONWORK, PIPES, ETC.

Iron houses also require frequent painting, not only in order to preserve the material, but to prevent the rust that forms if the ironwork is not kept coated with paint, from discoloring plants, walks, woodwork, and anything else that it may fall upon, with the drip. All ironwork that forms part of the greenhouse structure proper, should be of the same color as the woodwork. When iron tables are used they should be kept well painted, using some color of asphalt or Japan paint,— black asphalt being cheap and quite durable.

For the sake of the improved looks, to say nothing of increasing their durability, the heating pipes should also be painted. While asphalt will answer for this purpose, it is known that a larger amount of heat will be radiated from them if of a dull color, than if they are smooth and glossy, and the efficiency of the pipes will be increased by applying a mixture of lampblack and turpentine. The durability of the paint will be improved by using linseed oil, but it will have a glossy appearance, and if oil is used it should not form more than one-half of the mixture.

## SHADING.

In order to keep down the heat and prevent the burning of the foliage of the plants, it is desirable to, in some way, obstruct the entrance of the heat rays. For some classes of plants a permanent shading is desirable, and this can be secured by the use of fluted or rough plate glass. For most purposes, however, a temporary shading only is necessary, and some form of wash applied to the glass is commonly used to give this, when shading is necessary throughout the summer.

The application of lime or whitewash, either by means of a large brush or syringe, is a cheap way of shading the house, and is commonly used by commercial florists; but it is hardly satisfactory, as, when thick enough to keep out the heat rays, it obstructs too much of the light. One reason for this is, that if a coat of the proper consistency is given, it frequently peels off in spots, and when a second application is made, to cover these openings, it is too thick upon the other portions of the glass. This wash, too, has a glaring appearance, that is not pleasing to the eye.

Perhaps the most satisfactory shading is made by the use of either white lead or whiting, in gasoline. A very small amount of lead,—perhaps a teaspoonful,—will suffice for a gallon of gasoline, but the quantity of whiting required will be much larger. It will be best to make a thin preparation, and, if found to be too thin, more of the lead or whiting can be added. This wash can be put on in a fairly satisfactory manner with a syringe or small force pump, but it can be spread more evenly and with greater economy of material with a large brush, and, where the appearance is considered, this will be a better way. It is generally desirable to put on a thin coating early in the spring, and add a second one in May or June. If not put on too thick, the fall rains and frosts will loosen the shading, and it will disappear as winter comes on. If this does not take place soon enough, the roof can be wet down with a hose, and any surplus rubbed off with a stiff brush.

## TEMPORARY SHADING.

For orchid houses it is desirable to have a form of shading that can be regulated at pleasure. Some of the roller blinds answer well for this purpose, as they can be lowered on bright, sunny days, and drawn up at night, or in dull weather, to suit the needs of the plants. Cloth shades of light canvas or fine netting are less desirable, but answer very well. They can be used either outside or inside the house, and, if hung on curtain or awning fixtures, can be raised or lowered at pleasure.

When orchids are suspended from the sash bars, the shutters, canvas, netting or other material used for shading, must be placed above the glass, and, to allow a circulation of air above the roof, iron rods should be so arranged that the shading material will be supported at a height of twelve or fifteen inches. By means of ropes and pulleys, the awnings can be easily raised or lowered.

For shading cutting benches, there is nothing better than light frames covered with cotton cloth, although lath screens are very useful.

## CHAPTER XV.

### GREENHOUSE HEATING.

In our climate, most of the plants grown in greenhouses require artificial heat to be maintained from six to nine months of the year, in order that natural conditions may be secured for them. While some plants are not injured by exposure to thirty-two degrees, and thrive best at forty-five to fifty degrees, the so-called stove plants should have seventy degrees, or more, and to secure these temperatures in greenhouses various methods have been devised.

The crudest method is by slowly decomposing vegetable materials, and allowing the heat to radiate into the air; 2d. the Polmaise system, which consists in passing cold air over a hot iron surface, and directing it into ... as ... by burning wood or coal in a furnace, ... the gaseous products of combustion through the house in a brick or tile horizontal chimney, ... a flue which differs only in the method ... the heat, as in this is taken up by water ... water ... in the form of steam, or by ... circulation of the water itself.

... is only employed in hot beds and ... the ... known as the Polmaise system ... for greenhouse heating, although when combined with the flue, it is sometimes used.   In some sections of the country the flue is still made use of in heating small greenhouses, but by most florists steam or hot water is preferred.

Whatever the method of heating used, the average person would consider, in making a selection, the first cost and the durability, the economy of fuel and attendance, and the efficiency, both as concerns the amount and the regularity, of the heat supplied.   Among other things that would be taken into account, are the evenness with which the heat would be distributed, the length of time the systems will run without attention, and the effect of each upon plant growth.

## HEATING WITH HOT WATER.

This system was one of the first to be used for the heating of greenhouses in modern times, and it is claimed that the circulation of hot water, as a means of conveying heat, was used by the old Romans in warming their dwellings.   It went out of use, however, until 1777, when a Frenchman, Bonnemain, reintroduced it.   Ancient as the method is, the hot water heating systems of to-day are comparatively modern inventions, and bear little resemblance to those used even fifty years ago; in fact, the change has been so recent that many of the systems in use to-day are built on quite different principles from those constructed according to the latest ideas.

## HOT WATER IN THE EARLY DAYS.

The Romans are believed to have used bronze circulating pipes, and the first pipes used for heating greenhouses were of copper, and measured four to five inches in diameter.   The heaters used were also of copper, and generally resembled an open kettle resting upon a brick furnace.   From the kettle two four-inch pipes ran to the other end of the house, where they entered a copper reservoir (Fig. 3). The pipes were perfectly level, and one left the heater at the top, forming the flow, while the return entered at the bottom.

For thirty years previous to 1880, the usual method of heating greenhouses was similar to the one described above, except that closed cast-iron heaters were used, from which cast-iron pipes carried the water about the houses, ending in large open expansion tanks or distributing reservoirs.

### MODERN HOT WATER HEATING.

Modern heaters are made in hundreds of designs, and while each is generally claimed, by its inventor, to surpass all others, it is a hard matter to decide which one is really best. They are made of both cast and wrought iron (small ones may be made of copper, zinc, etc.), and here, at once, arises a dispute as to the merits of the two materials.

The wrought iron is more likely to rust and, during the long summer months, when they stand unused in the damp greenhouse stoke-holes, they often suffer severe injury. The wrought iron is, also, more injured than cast-iron, by the sulphurous and other gases of combustion and, for these reasons, it is claimed by some that cast-iron boilers will last much longer than those of wrought iron. This has certainly been the case with some heaters, but it has been due, in part, to the fact that many heaters have been made of common gas pipe, instead of the double strength pipe which should be used. When this thin pipe is threaded, and the threads are not made in, the surface exposed is quickly eaten through. When no pipes smaller than one and one-fourth inch are used, and these are double strength boiler flue pipes, the durability of the wrought-iron heaters will be increased.

## POINTS FOR A HOT WATER HEATER.

Aside from durability, simplicity, and compactness of construction, the following points in the make-up of the heater should be considered: 1. The amount and arrangement of the direct heating (fire) surface, and its proper adjustment to the grate area. 2. The arrangement of the water sections, or tubes, and the circulation of the water in the heater. 3. Ease of cleaning the flues, and the arrangements for shaking, dumping, removing the ashes, regulating the draft, etc. 4. The character of the joints, and the ease with which leaks can be repaired, and breaks mended.

If the first and second requirements are met, we may have a heater that is efficient and economical of fuel, but the points noted in the third have much to do with the ease of firing and caring for the heater, while those in the fourth will be desirable in case leaks occur.

### 1. ARRANGEMENT OF THE FIRE SURFACE.

It is well known that a surface arranged at right angles to the fire is nearly twice as efficient as one that is parallel to it. Unless this can be secured, it necessitates a corresponding increase of the area of fire surface, which will not only add to the cost of the boiler, but will render it more cumbersome, and increase the amount of circulation of water in the heater. When the arrangement is such that horizontal surfaces cannot be secured over the firepot, the same effect can be, in part, obtained, if the direction of the draft is such that the flames are drawn at right angles towards perpendicular tubes. When this can be brought about, it affords very effective heating surface, and is not objectionable; on the contrary, it is desirable to so arrange the draft and flues, that the products of combustion are carried in as indirect a course as is possible, and yet secure a proper draft for combustion, removal of smoke, etc. By doing this, and by repeatedly bringing this heated air in contact with the water sections, we can finally lower the temperature down approximately to that of the water. The nearer we approach this, the greater economy shall we find in the heater.

While it is of importance that heaters have ample grate areas and a good draft, the amount and arrangement of the fire surface is of equal importance. To obtain the best results, the grate area and fire surface should be carefully adjusted; but for this no general rule can be given, as some heaters have their surface so nicely arranged that the heat liberated upon one square foot of grate area can be taken up by fifteen square feet of heating surface, while in other heaters thirty-five or forty feet of fire surface will be insufficient. In a general way, a square foot of grate surface will supply two hundred and fifty square feet of radiating surface, but, as a rule, it will be more economical if two hundred square feet of radiating surface is taken as the limit.

## 2. ARRANGEMENT OF THE WATER SECTIONS AND TUBES.

The arrangement of the fire surface will, of course, determine the position of the water in the sections and tubes, but will not, necessarily, regulate the direction of the flow, the amount of water, etc. The circulation of the water in the ordinary heaters is vertical, horizontal, in drop tubes, or a combination of two, or even all three of these ways.

The circulation in the heater should be as short as possible, and it is better to have the water spread out in thin sheets, and with the arrangement such that the water is divided into a number of portions, each of which makes a single short circulation, than it is to have it so that the mass of water that flows through the heater warmed by convection, or compelled to pass in a zigzag course through a number of different tubes and sections. In this way, too, the friction will be decreased and the circulation improved.

So far as circulation goes, the vertical tube tends to reduce friction, and to this extent it is desirable. On the other hand, the friction produced by one circuit of the water in a horizontal section is so slight that it is often more than counterbalanced by the increased efficiency of the horizontal fire surface.

The drop tubes used in many boilers present a very good fire surface, as the ends are directly over the grate, and, as the water circulation is vertical, they form a very effective portion of the heater. When large tubes are used there is little danger of their filling up with sediment, and the principal objection that can be urged against them is that the water cannot be drawn off from them.

Another thing that it is desirable to secure, if possible, is the bringing of the products of combustion, as they are about to leave the heater, in contact with tubes or sections containing the return water. It can be readily seen that water at 175° coming back in the returns, can still take up heat from gases that have been in contact with iron surfaces that are 200° or more. In this way considerable heat will be saved that would otherwise pass up the smoke pipe.

### 3.  ARRANGEMENTS FOR CLEANING AND FIRING.

It is self-evident that anything that adds to the convenience of a heater will be desirable, and the matter of shaking, dumping, and regulating of drafts should be considered. Of especial importance, however, is the matter of cleaning the flues. Unless there is a great loss of heat, a heater cannot be made in which there will not be, in some portion, an accumulation of soot, and if this is upon any of the heating surfaces it should be frequently removed. A heater in which the flues cannot be kept clean is of little value, and the greater the ease with which it can be done, the better. If the surface requiring cleaning is small and easily cleaned, the actual trouble would be very slight, and although the flues of some heaters are practically self-cleaning, their heating surface may be less effective, which would more than counterbalance the cleaning required by the other.

### 4.   SIMPLICITY OF CONSTRUCTION.

Many heaters are quite intricate in their construction, and the different parts are fastened with screw joints, or, as is more common, the joints are packed and the parts are drawn together with bolts. Everything else being equal the fewer joints there are, the less chance there will be of leaks, and in selecting a heater this should be considered, as well as the character and location of the joints. The screw joint is perhaps the surest, but it has one ojection, particularly in wrought-iron heaters, as the threads tend to increase the corroding influence of the sulphur gases. Packed joints are fairly reliable, but it is desirable that they be easy of access, and so arranged that they can be repacked should serious leaks occur. Some of the sectional heaters are so constructed that, if one section is broken in any way, it can be cut out of the circulation, and the heater can then be used without it until the section is mended or a new one procured. In case the section has to be replaced by another, the arrangement should be such that the change can be readily made.

While all of the points enumerated above are deemed desirable in a heater, it can still be of great value if it does not possess one or more of them; but in selecting a heater, while the fact that one or more important features were lacking might not prevent its being chosen, it would be well to take the one which comes the nearest to possessing them all.

# CHAPTER XVI.

## PIPES AND PIPING.

In the old styles of hot water plants, the pipes were of four-inch cast-iron, put together with shoulders and packed joints, and with large expansion and distributing tanks at the ends of the runs, and at the points where the branches left the main lines. In the modern system two-inch pipe is the largest used for the coils, while one and one-fourth inch and one and one-half inch are preferred for short runs.

Some of the advantages of the modern system may be stated as follows: The lengths of pipe are from two to nearly four times as long, and can be screwed together instead of having to pack the numerous joints; there is less chance of leaky joints or of cracked pipes; although the cast-iron four-inch pipe has only twice the radiating surface, it is necessary to provide four times the amount of water for the circulation that the two-inch contains; on account of the size and weight of the four-inch pipe it is necessary to have them low down under the benches but little above the level of the heater, while the small wrought-iron pipe can be carried in the very angle of the ridge if desired, and thus a far more rapid circulation can be maintained than with large pipe; the large pipe carrying a large quantity of water and giving a slow circulation, is at a much lower temperature, particularly on the returns, and a smaller radiating surface will suffice when small pipe is used, so that some florists count a two-inch wrought-iron pipe equivalent in heating capacity to a four-inch cast-iron pipe; from the

7                               97

large pipe the amount of heat given off on bright, sunny days when it is not needed, will be from two to four times as much as from small ones, and this will necessitate increased ventilation and perhaps cause serious injury from drafts of cold air, to say nothing of the loss of heat ; finally, in addition to the points enumerated above, the small pipes will give a much more economical circulation than the large ones, they can be carried to a much greater distance, and the heat will be far more even.

It has been claimed that the large piping is safer to use, as it will hold the heat longer. This is undoubtedly true, if the fire is allowed to go out ; but, with a well-arranged system, a regular, even temperature can be maintained with small pipes for ten to twelve hours, on mild winter nights, and seven to eight hours on severe nights, which is as long as it is desirable for the houses to go without attention.

The term " upward pressure " is often used in speaking of hot water circulation ; but although it is a convenient one to use, it really has nothing to do with the circulation, as there is no pressure of the kind exerted. Hot water has a *downward* pressure equal to its own weight, and the only reason for circulating is that the weight of the hot water is more than balanced by the weight of the cooler water in the pipe, and it passes upward, pressed out of the way by the heavier cool water which pushes into its place. The same thing can be seen in a kettle of water where the water in the center is warmed and is pushed to the top, while the cool water from above takes its place. The method of circulating in a hot water apparatus can be best understood by reference to one of the old styles, as shown in Fig. 3 (a represents the heater, b the expansion tank, and c the flow and return pipe). Let us suppose that a fire is built under the boiler, and that the water contained therein

becomes warm, the same as in an ordinary kettle. It is
known that water when warmed from 39° to 212° increases in bulk one twenty-fourth. If the water in *a*
is warmed up to the boiling point it has decreased in
weight, *per cubic inch*, one twenty-fourth. The water
then in *e* is one twenty-fourth heavier than in *a*, and, to
establish an equilibrium, the water in *e* will pass along
the lower pipe to *a*, crowding the lighter water into the
upper pipe.

If the heat is continued, other particles are set in
motion the same way, and the rapidity of the circulation
will increase until it is balanced by the friction. The
circulating force is governed by the comparative weight
of the warm water in the different parts of the system.
The pressure of the water varies with the height of the
columns, as well as the temperature of the water. If
the height of the columns is increased, the difference
between the weights of the two columns will be increased
in about the same ratio, and as this *difference* in weight
is what causes water to circulate, the reason for the
success of the overhead system of piping can be readily
seen. The same effect could, however, be secured were
it convenient to do so by lowering the heater.

There has been considerable discussion for many
years as to the best way of running the pipes, but even
now very few persons agree as to the proper method of
arranging them.

### HOW SHALL THE PIPES SLOPE?

Among the various methods are the "up-hill,"
"down-hill," and "level," and these are shown in Fig. 37.
1, 2, and 3; the last, however, is not desirable when
small pipes are used. In each case the height of the flow
pipe at the point where it starts to make a circuit of the
house is six feet above the bottom of the boiler, but in
the first case the pipe rises one foot in passing through

the house; in the second it falls a foot, and in the third
it does not change its level. In changing direction at
the farther end, a foot fall is made by each pipe, and on
the return a fall of one foot is made by the first and
second systems, while the third remains at the same
height until it has nearly reached the heater, when it
drops to the level of the bottom of the heater.

The "pressure" is determined by the relative weight
of the water on either side of the highest point of the

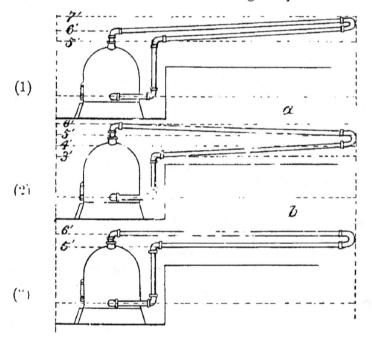

FIG. 57. THE SLOPE OF THE PIPES.

system, and it would have been a fairer comparison had
the flow pipe for the "up-hill" system left the heater at
a height of five feet rather than six, as the highest point
of each system would then have been six feet above the
bottom of the boiler. When the size and the length of
the pipes, the connections, etc., are the same, the system

that is arranged to give the greatest difference between
the weights of the water on either side of the highest
point, will have the best circulation, everything else being
equal.  For convenience let us consider that in each
system the "flow" pipe extends from the heater to the
highest point of the piping, and the returns extend from
that point back to the boiler, entering at the bottom.
To secure a good circulation, the water in the flow pipe
should be as light (hot) as possible, and, that it may not
be subjected to cooling influences, the pipe should be as
short as possible.  If the flow pipe is short, then the
highest point in the system must be near the heater.

In order that there may be a difference in the weight
of the two columns of water, that in the return pipes
should be as heavy (cool) as possible, and this can be best
secured, everything else being equal, if the distance is
considerable.  Turning to the illustration we shall see
that the highest point in (2) and (3) is directly over the
heater, while in (1) it is at the extreme end of the system :
the return in (1) is only one-half as long as in (2) and (3),
so that the cooling will be only about one-half as great.

As we wish to keep the water in the pipe between
the heater and the highest point of the system from
cooling, a large pipe should there be used, while, as it
is desirable to cool off the water in the returns, the
remainder of the system should be of small pipe.  Considering the average temperature of the water in the flow
pipe to be 200° F., and that in the returns 170 , a cubic
foot of the latter will be one-eightieth heavier than a
cubic foot of water at 200 .  If a pipe one foot high
contains one pound of water at 200°, the same pipe will
hold 1.0125 pounds of water at 170°, and were the two
united there would be a "pressure" of .0125 of a pound.
Were each pipe ten feet long, the water in one would
weigh ten pounds and in the other 10.125 pounds, and
there would be a pressure of an eighth of a pound,

while at eighty feet there would be one pound pressure, or eighty times as much as at one foot.

It will thus be seen that the pressure increases with the height of the columns. As we wish to have as much weight as possible in the returns, they should be brought back to a point near the heater at as near a level as possible, and at the greatest convenient height. As just shown, if a flow pipe is filled with water at 200° and we consider it to weigh one pound for each foot of pipe, there will be a pressure of .125 of a pound from a return filled with water at 170 F., if the pipes are each ten feet high. This would be the pressure then in a hot water apparatus under the above conditions, *i. e.*, the flow pipe rises ten feet above the heater, and passing through the house returns to a point over the heater, without changing its height.

On the other hand, suppose that after passing to the farther end of the house it drops perpendicularly five feet and returns at that level to the heater. There will then be a head of five feet where the water has a temperature of 185 and of five feet when it has cooled down to 170°, and we should have

$$(185) \ 1.0057 \times 5 = 5.0285$$
$$(170') \ 1.0125 \times 5 = 5.0625$$

$$10.0910 - 10. = .0910 \text{ lbs.}$$

As compared with a pressure of 0.125 of a pound when the return was ten feet high, this shows a pressure of only .0910 of a pound, or a loss of .034 of a pound as compared with the other method. In order to allow the air to escape, the pipes cannot be carried on a level, and hence as the next best method they should be given a gradual fall throughout their length, taking pains to keep them as high as possible in order to secure pressure. While it will be best to have long straight runs, with the same slope throughout, if necessary there may be vari-

ous changes in the level of the pipes, and circulation can
still be kept up, provided the necessary vents for removing the air are provided, and the pressure, as explained
above, is sufficient to overcome all friction.

In some cases the down-hill piping (2), Fig. 57,
cannot be used to advantage, and the up-hill system (1),
will serve the purpose. With the slight difference in
level that can be secured in greenhouse heating, particularly if the pipes are under the bench, the difference in
efficiency will hardly be noticeable. Theoretically, the
level piping, as in Fig. 57 (3), gives the greatest pressure ;
the down-hill system comes next, and the up-hill piping
gives the least pressure, although, as piped in Fig. 57 (1),
the pressure with a coil seven feet high is about the same
as with the level piping in a coil six feet high.

## OVER VS. UNDER BENCH PIPING.

It is unquestionable that if all the pipes are above
the benches, the circulation will be better than if all, or
even a part of them, are below. If one or two pipes are
placed upon each plate, and the others near the purlins,
the amount of light obstructed will be comparatively
small ; the heat, however, will not be as well distributed
as if the pipes are spread out below the benches. Experi-
ments to test the matter have shown little if any difference in the results, whether all of the pipes are above or
below the benches, but with the feed pipe above the
bench and the coils below, there is an improvement in
circulation, and fully as good if not better growth of
plants as when all are below, and this plan should be
used whenever practicable.

The principal benefits from overhead piping are (1)
the melting of snow and ice on the roof, (2) taking the
chill from cold drafts of air and (3) drying off the
plants after syringing, all of which will be largely done
by the overhead main. The overhead system, moreover,

carries the coils higher than they are in the under-bench system and a more rapid circulation is secured.

Some houses have been piped with the flow pipes under the side benches, and the returns above the benches. While this gives a good circulation, better results can be obtained if the pipes are placed in the same way except that the overhead pipes are attached to the flow pipe from the boiler, and those under the benches to the returns. The returns under the bench can either be of the same size as those above, or of a larger size.

## CHAPTER XVII.

### SIZE AND AMOUNT OF PIPE.

The size of pipe best suited for the coils depends upon the length as well as upon the height of the coils. Considered as radiating surface only, one-inch pipe would be preferable, but, except for very short runs, since the friction increases as the size of the pipe decreases, a larger size should generally be used. Inch and one-quarter pipe can be used to advantage in coils not over forty feet long, and if the height is sufficient, the length may be considerably increased. For coils up to seventy-five feet in length, one and one-half inch pipe will be entirely satisfactory. Two-inch pipe will work well up to one hundred and fifty feet, but it is better in all houses over one hundred feet long to use two or more short coils of one and one-half inch pipe. In this way by having proper flow and return pipes connected with the coils, houses three hundred feet long can be heated with hot water.

It it can be so arranged, however, it will be better to have houses one hundred and fifty feet long on either side of a potting shed and connecting passage, which will really make houses three hundred feet long while, with the heater located at the center, they will be only one hundred and fifty feet in length so far as the heating apparatus is concerned.

## SIZE FOR MAINS.

In determining the size for the feed pipes the length of the house and the height that the coils will have, should be considered, as well as the number of square feet of radiation to be supplied. For houses of average length, and with the average height of the coils six feet above the bottom of the heater,

2 inch pipe will supply 200 to 300 square feet of radiation.
3    "    "    "    600 to 800    "    "    "
3½    "    "    "    800 to 1000    "    "    "

With long coils and light pressure these figures will need to be slightly reduced, but if the runs are short and the coils elevated they may be increased fifty per cent.

## TO ESTIMATE RADIATION.

In computing the amount of radiation required for a house, the climate, exposure, construction of the house, and the amount of exposed wall surface should be considered, as well as the temperature to be maintained. When the walls are high or poorly built, or if the roof is not tight, allowance should be made for it by adding to the glass surface one foot, for every five feet of exposed wall, and a corresponding increase for the poor glazing.

In most parts of the country we can reckon that

1 square foot of pipe will heat to 40  4½ square feet of glass.
1    "    "    "    "    "    50  4    "    "    "
1    "    "    "    "    "    55  3½    "    "    "
1    "    "    "    "    "    60  3    "    "    "
1    "    "    "    "    "    65  2½    "    "    "
1    "    "    "    "    "    70  2    "    "    "

While a slightly higher estimate would be safe, it is economy to have an abundant radiation, especially for tropical houses.

In estimating the surface of wrought-iron pipe,

1 inch pipe is reckoned at .344 square foot per linear foot.
1¼ "        "        " .434 "    "    "    "
1½ "        "        " .497 "    "    "    "
2 "        "        " .621 "    "    "    "

### PIPING THE HOUSES.

In arranging the heating pipes we can place them all under the benches, have them all above the benches, or the flow pipes may be over the tables and the returns underneath. If the under-bench system is used, the arrangement shown in Fig. 58 is a good one. When the house is one hundred and fifty feet long a three-inch

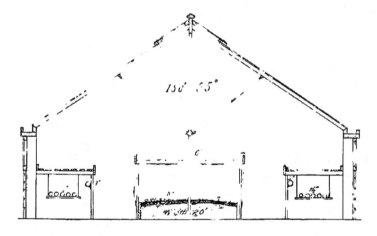

FIG. 58.   UNDER BENCH PIPING, WIDE EVEN SPAN HOUSE.

flow pipe will be needed for each of the side benches. (A two-inch pipe will answer for a house one hundred feet long.) When the coils are of one and one-half inch pipe there should be two, each seventy-five feet long, per each side or three of fifty feet each. With two coils

the mains should enter the house close to the bottom of
the bench, and falling at the rate of one inch in twenty
feet should pass to the middle of the house, where a two-
inch branch should be taken off at the side by means of
a hot water T; the main should be continued by means
of a two-inch pipe to the farther end of the house, where
it should connect with the coil.

In a house one hundred feet long the two-inch feed
pipe should be run in the same way as the above, except

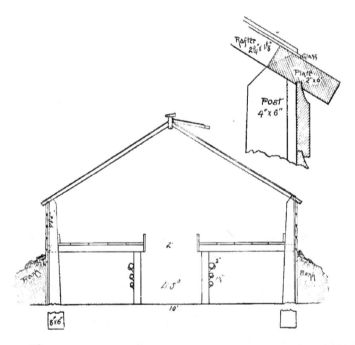

FIG. 59.  UNDER BENCH PIPING FOR NARROW EVEN
SPAN HOUSE.

that the branches should be of one and one-half inch
pipe.  If desired, however, the coil can be of two-inch
pipe, in one piece.  The return pipes should fall towards
the heater at the rate of one inch in ten or twelve feet.
If the bench is too low to admit of a proper fall of the

low pipe from, and of the return towards the heater, the
flow may be given a gradual rise to the farther end, and
thus a fall can be secured for the returns.

Another method that will be preferable to either of
the above when the bench is high enough to give a fall,
or if the return can be placed below the level of the
walk, is to attach the feed pipes to the coils at the end
nearest the heater. The coils can then be given a

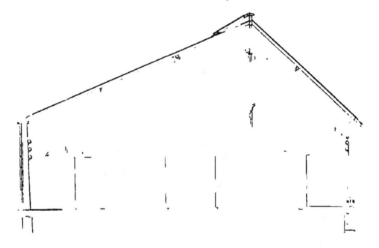

FIG. 60. OVERHEAD PIPING SHORT SPAN TO THE
SOUTH HOUSE.

fall from the heater, and the returns can come back
underneath. The arrangement of the pipes in a narrow
even span house to be heated to forty-five degrees is
shown in Fig. 59.

## OVERHEAD PIPING.

In carnation and other cool houses where the amount
of pipe is so small that it can be carried upon the plate
and purlins, and particularly in a short-span to the south
house, the overhead piping will perhaps be desirable
The feed pipe can be carried upon the ridge posts

(Fig. 60), and the returns arranged as shown in the sketch, or the flows, as several small pipes, may be above, and the returns, as one or two large ones, below.

## COMBINED OVERHEAD AND UNDER-BENCH PIPING.

For most commercial establishments the above arrangement will be preferable to having all the pipes above, or all under the benches. One method is illustrated in Fig. 61, in which the main is carried near the ridge, and the returns in vertical coils upon the bench legs. In the north-side propagating house, all of the

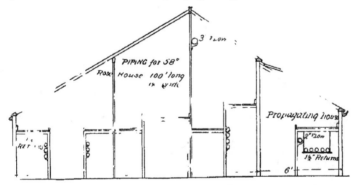

FIG. 61. COMBINED OVERHEAD AND UNDER-BENCH PIPING.

pipes are under the bench. For an even-span house one hundred feet long and twenty feet wide to be carried at sixty-five degrees, the arrangement illustrated in Fig. 62 will give good results.

With a good fall, one and one-half inch pipes in the coil can be used for a long run, but it will generally be better to make two coils on a side, each fifty feet long. At the middle of the house, feed pipes (one and one-half inch) can be taken off to feed the first coils, and the main can be extended as a two-inch pipe to the end of the house, where branch pipes can be connected with the other coils. Fig. 63 illustrates a method of piping a forcing house one hundred and fifty feet long and twenty feet wide that is to be kept at sixty-five degrees. The coils, as in Fig. 62, are placed horizontally, which probably makes them more efficient than when they are arranged vertically. Fig 81 shows a method of arranging the heating pipes under the benches that can be used when there is a bench across the farther end of the house. As the main is somewhat elevated, it has some of the merits of the combined system. It will be noticed that although Figs. 52, 63, and 81 are designed to illus-

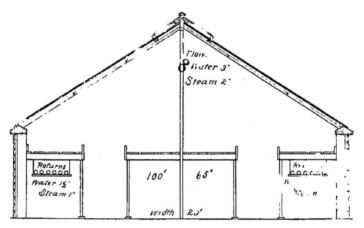

FIG. 62.　COMBINED PIPING, EVEN SPAN HOUSE, 100×20 FEET; 65 DEGREES.

trate methods of piping, they also show various ways of building the walls, ventilators, posts and braces, benches, etc.

### PIPING NARROW HOUSES.

In low and narrow houses the same methods of piping can be used, making proper reductions in the number of the pipes in the coils, and in the size of the main. Thus, for a span-roof house twelve feet wide and one hundred feet long, to be kept at fifty degrees, a two-inch pipe at the ridge will feed three one and one-half inch returns under each bench, or, if the under-bench system is to be used, a two-inch flow on each side will feed two one and one-half inch returns. Still another method would be to use two one and one-half inch flow pipes under each bench, to feed the same number of returns.

The heating surface will be most efficient if it is distributed evenly under the benches, but as they will then occupy considerable space that could be occupied by mushroom beds (Fig. 58) or for other purposes, it will generally be preferable to group them under the side benches.

In building the coils, cast-iron headers, or, better, manifolds built up of Ts and nipples, can be used at one

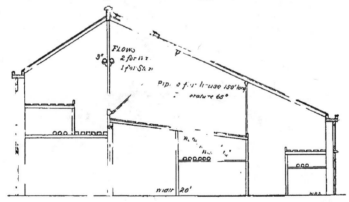

FIG. 63.  COMBINED PIPING FOR FORCING HOUSE,
150×20 FEET ; 65 DEGREES.

end of straight runs, but on account of danger from the unequal expansion of the pipes, a mitre should be arranged if headers are to be used at both ends. The coils can be carried across the end of the house and end in headers at the center, to which branches from the feed pipe can be attached, as in Fig. 64.

The overhead mains can be fastened to the posts by means of cast-iron brackets, while under the bench they can be supported by pipe hooks upon the bench legs. The vertical coils can be fastened in the same way, and the horizontal coils can be supported as in Fig. 63, or upon pieces of gas pipe suspended from the bench cross-bearers by iron hooks, as seen in Figs. 58 and 62.

### VALVES AND EXPANSION TANK.

In order to regulate the flow of the water through the coils, there should be angle or gate valves upon the returns, or upon the branch feed pipes, and if it is desirable to arrange the house so that heat can be shut off for

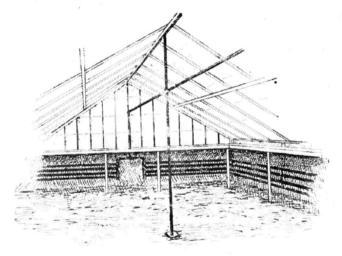

FIG. 64.    ARRANGEMENT OF THE COILS.

a portion of the winter, there should be valves on both the flow and return mains where they enter the house, with a draw-off cock so that the water can be removed.

The expansion tank should be raised as high above the mains as possible, and connected with it (at the highest point, in case the piping is down hill), by a pipe from one to two inches in diameter, according to the extent of the system.  The elevation of the tank in no way affects the circulation, as it merely raises the boiling point of the water. The tank may be of galvanized iron, without a cover, with the expansion pipe connected with the bottom, and an overflow pipe attached about two-thirds the way up the side, or a riveted boiler-iron closed tank with the same connections and a water gauge on the side.

## HOT WATER UNDER-PRESSURE.

When the down-hill system is used, many florists combine a closed expansion tank with it. The tank should be of boiler-iron with top and bottom securely riveted on; thus far it does not differ from those commonly used in the open system. The only difference is that there is no vent in the top of the tank, and that there is a safety valve on the overflow, which, as in the open tank, carries the waste water to a drain.

The closed tank has the same effect as does the elevated one, and merely raises the point to which the water can be heated without forming steam. One advantage of this is that, when water is carried at 220°, much less heating pipe is required than when it is only 160°, but a serious objection is that it now has one of the faults of steam as, at this temperature, more heat will be carried up the chimney with the products of combustion, than when the water is 180°. It is an excellent plan to have the house supplied with sufficient radiating surface to maintain the required temperature in the average winter weather with an open tank, but to have the system provided with a safety-valve, which could be thrown on when it became necessary, in order to keep up the temperature when the thermometer goes down below zero.

Unless the down-hill system with the tank at the highest point, is used, air vents should be provided, wherever the pipe takes an upward turn, at the highest points. Air cocks can be used, or quarter-inch gas pipes

8

carried above the level of the expansion pipe will answer, and are preferable if the tank is not much elevated.

### CHANGES OF DIRECTION AND LEVEL.

If for any reason it becomes necessary to change the direction or the level of the pipes, it should be the least amount possible, and, in doing it, 45° ells, reducing tees, etc., should be used. If necessary the main pipes can be changed from their course and run over or under an obstruction, and they can even be carried below the level of the heater, but it should not be done unless there is an abundant pressure. In all downward changes of level, the force required to bring the water back to its original level will be in proportion to the cooling that takes place between the fall and rise. Every change in direction increases the friction and decreases the flow.

### CONNECTING THE DIFFERENT SYSTEMS.

If there is a range of houses to be connected with one system, the arrangements will need to be such as will suit the conditions. If, as is very convenient, the houses have a common head house, the heater can be situated in the center, and the feed mains can be carried along the wall of the head house, just above the doors that open into the greenhouses, and the branches can be taken off from this. The returns can be connected in much the same way, but the return main will be below the level of the coils.

### FOUR-INCH HEATING PIPES.

Although an expensive method of piping and heating greenhouses, many florists prefer to use four-inch pipes. The pipes must all be under the benches, upon substantial brick piers. If air vents are provided at frequent intervals the pipes may be level; otherwise a slight down-hill arrangement will be preferable. The joints should be packed firmly with oakum or tarred rope, with iron cement or Portland cement between the layers, and at the outer edge of the joint. If the runs are long the pipes should rest on gas-pipe rollers, that expansion and contraction may not break the joints.

### PIPING IN GENERAL.

The same general rules as to the arrangement of the pipes apply to all kinds of houses, and will only need to be slightly modified to suit the various conditions. For large conservatories, particularly if the center of the house is filled with plants growing in the ground, it will be necessary to have all of the pipes arranged in stacks along the sides, or, if desired, cast-iron radiators can be used.

## CHAPTER XVIII.

### HOT WATER HEATERS.

In the illustrations of heaters presented, the selection was made with the idea of showing the construction of different types rather than of advocating the use of

any particular heater. While we can recommend most of these from our own experience, there are other heaters that may be fully as effective.

The Carmody heater (Fig. 65) is selected as the type of the *vertical* sectional heaters. They are of a very durable construction and, like others of the class, possess the important

FIG. 65. CARMODY HEATER.

advantage of permitting the addition of other sections, should it at any time become necessary.   The water cir-

FIG. 66.   HITCHINGS' HEATER.

culation, for the most part, is vertical, and the flues are arranged to give an effective heating surface.

FIG. 67.   WEATHERED CONICAL HEATER.

In some respects similar to the Carmody, but differing in being non-sectional, are such well-known heaters as the Hitchings, Smith, and Weathered. They differ principally in the arrangement of their fire surfaces. In the Hitchings' corrugated heater, of which a longitudinal section is shown in Fig. 66, the fire surface is increased by means of rounded corrugations, while, in the Smith, the corrugations are replaced by quite deep square cells.

Few heaters, with equal grate areas, will surpass those of this class in the amount of heat they will furnish, or in the economy of their coal consumption. For cool houses, or when carrying about two-thirds of their full radiation, they give very satisfactory results. While their fire surface is very effectively arranged, it is rather small for the grate area, and, in case the heaters are working up to their full capacity, in very cold weather, there will be a considerable loss of heat through the smoke flue. With these heaters, as with others, it is economy to select one that is a size larger than would be really necessary. For small greenhouses the Weathered Conical Heater (Fig. 67) or others of similar construction will be found quite powerful and efficient.

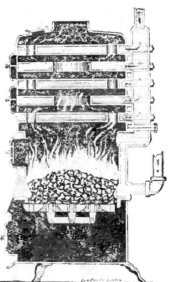

FIG. 68. SPENCE HEATER.

In Fig. 68 we present the Spence as showing the general form of the *horizontal* sectional heaters. The fire-pot is surrounded with a water jacket so that a fire-brick lining is not required. The sections are arranged parallel, one above the other, over the fire-pot. The construction of the flues is such as brings the products of combustion repeatedly in contact with the heating surface, and if this is not sufficient to absorb all of the heat, the difficulty can be corrected by the addition of more sections. The first, third, and fifth sections are shown in Fig. 69, and the second and fourth in Fig 70. The water is spread out in a thin sheet in the sections and cannot enter the feed pipes until it has made a complete circuit of one section. By the arrangement of the water column in the rear, the water having passed through one section, is prevented from entering another. The points claimed for this heater are, economy of fuel, perfect and rapid circulation, readiness of cleaning, few and tight joints, and, in case of leakage or breaks, the readi-

FIG. 69.   SPENCE HEATER.     FIG. 70.   SPENCE HEATER.
(1st, 3d and 5th Sections.)     (2d and 4th Sections.)

ness with which repairs can be made. The Gurney, (No. 300 Series) and the Palace King have much the same construction, and are, very likely, fully as efficient.

The Furman heaters may be taken as the type of the drop tube patterns. They are used either for steam or hot water, the small sizes generally being portable (Fig. 71), and the large ones brick-set (Fig. 72). The Furman hot-water heater is constructed on principles almost exactly the reverse of those found in the Spence. It is not sectional in the small sizes, and yet, as the parts are screwed together, they can be taken apart if any of the tubes are broken. The cast-iron screw joints will prob-

ably never leak. It will be noted that the heating surface consists of oval drop tubes, with a diaphragm in the
center, so arranged that the water passes down on one
side and up the other. As previously stated, the vertical
circulation, as secured in the Furman, is the correct one,
as there is less friction than in horizontal tubes or sections. It gives a very rapid circulation, which is of
importance in taking up the heat and, also, in giving it
off in the coils. A happy illustration of the effect of a
rapid circulation upon the amount of heat taken up by
the water, is the wind blow-
ing over a muddy road, the
faster it moves, the more
water it takes up. The
faster the water travels past
the fire, the more heat will
it absorb, and the less will
pass up the smoke pipe.

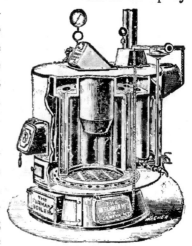

In the Furman, the at-
tempt is made to secure the
direct action of the fire upon
the tubes, by means of lateral draft between them;
this also tends to secure
perfect combustion to the
very edge of the fire-pot.

FIG. 71.   FURMAN PORT-
ABLE HEATER (*Steam*).

From their shape and arrangement, very little cleaning
of the flues is necessary.

Although constructed upon a general plan exactly
opposite that of the Spence, from our trial of the two
heaters here, it is not possible to decide which is the
correct one: if anything, however, the Spence is more
economical of fuel.

Where the size of the plant will warrant, it is better
to have two heaters in a battery, than one very large one,
and where two large heaters will do the work *on a pinch,*

it is desirable to have a third one to fall back on in severe weather, or in case of an accident to one of the others. If a single heater large enough to do the work in the most severe weather is used, it will be twice as large as will be required in the mild weather of Spring and Fall, but, by having two heaters in a battery, one or both can be used as may be necessary.

For use in small conservatories, there are many forms of portable hot water heaters, of which that made by Hitchings & Co., shown in Fig. 73, may be taken as a sample. Most of them have coal magazines, and run for eight to ten hours without attention. From the

FIG. 72. FURMAN BRICK-SET HEATER (*Steam*).

simplicity of the hot water apparatus as first used, it will be seen that good results can be obtained from almost any kind of a heater that provides for a proper connection. A simple can of copper, zinc, or galvanized iron, resting over an oil stove, will provide heat for a small conservatory, but if some arrangement can be made for increasing the heating surface, better results will be obtained.

THE SIZE OF HEATER TO USE.

Having determined upon the kind of heater to use, the size to obtain is of considerable importance. All greenhouse heaters are rated by the manufacturers as equal to supplying a certain number of square feet of radiation. Although most of them will do what is claimed for them *at a pinch*, it will be at the expense of an excessive amount of fuel and labor. The most economical results with hot water can only be obtained with a thin, slow fire in a large fire box, and as a rule it will

be well to deduct at least twenty-five per cent. from the manufacturers' rating in estimating the capacity of a heater.

The comparative area of grate and fire surface in heaters varies with their arrangement to such an extent that, provided it is ample to absorb the heat produced by the combustion, the latter may be left out of the question for the present. Basing the required grate area upon the number of square feet of radiating surface, it has been stated that for economy the ratio of one to two hundred should not be exceeded. With large

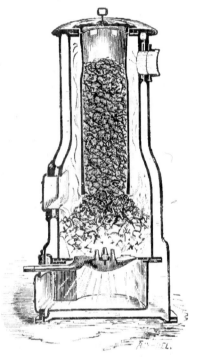

FIG. 73. HITCHINGS' BASE BURNING HEATER.

heaters this should suffice, provided the radiation itself was ample, but in small establishments, with less than one thousand feet of radiating surface, the proportion of one square foot of grate surface to one hundred and fifty square feet of radiating surface will be none too much for the economical consumption of fuel. In establishments where cheap fuel is used and a night fireman employed, one square foot of grate surface will burn enough fuel, with a good draft, to supply two hundred and fifty square feet of radiation.

Of course the ratio of radiating and glass surface must be based, in addition to the temperature to be maintained, upon the climate, the exposure, the construction of the house, etc., but, as a rule, the average temperature in a greenhouse may be taken as fifty degrees, and one foot of radiating surface will heat about four square feet of glass. For an establishment then of 2,400 square feet of glass, 600 square feet of radiating surface will be necessary, and a heater with a grate containing four square feet will be required. If it contains 8,000 square feet of glass, 2,000 square feet of radiation and ten square feet of grate surface will be necessary, and for 16,000 square feet of glass the radiation and grate surfaces will be respectively 4,000 and sixteen square feet. In the first two cases the fires can be left at night without attention for eight hours in zero weather, but would require stoking once in three or four hours when the grate surface is as small as given in the last example.

# CHAPTER XIX.

## STEAM HEATING.

With the wonderful growth of commercial floral establishments during the past ten years, a need arose for something more efficient and applicable to larger houses than the old-fashioned flue, or the hot water system with four-inch pipes, and it was found in the modern steam greenhouse-heating plants. In a general way, the same rules and method of piping would answer here as were given for hot water.

In steam heating we have the choice of two methods, high or low pressure. In the first it is preferable to use wrought-iron boilers rather than the average one of cast-iron, although some cast-iron tubular boilers are claimed by the inventors to withstand higher pressures than those of wrought-iron. This method of heating is particularly applicable in large plants with more than 12,000 square feet of glass, where a regular night fireman can be employed. The principal arguments in its favor are that less radiating surface is required than with low pressure steam or with water, and that steam can be carried to considerable distances, thus centralizing the boilers, and enabling the most extensive ranges of house to be heated from one boiler-room. For small plants the low pressure system, carrying a maximum of five pounds pressure, and generally not over two pounds, is preferable.

### STEAM BOILERS AND THEIR LOCATION.

Some of the horizontal tubular boilers are generally used and give general satisfaction. For low pressure

123

there are dozens of cast-iron boilers, each of which has
points that, if we can believe the inventors, makes *his*
the *best*; really, however, the difference in their real
efficiency is very slight. For small houses the locomo-
tive boiler seems to be a cheap and economical heater.
They are also used with success for hot water. As in
the hot water heaters, the requirements for a good steam
boiler are ample grate surface, good draft, and a fire
surface at right angles to the draft, with the flues so
arranged as to absorb the greatest possible amount of
heat.

In locating the boiler, pains should be taken to have
it low enough so that the water level will be at least two
feet below the lowest heating pipe, but if this is not possible without sinking the boiler in dark, poorly drained
pits, steam traps can be used, particularly with high
pressure, that will remove the water from the return,
and lift it to the water level of the boiler; with low
pressure they work more slowly and are less satisfactory.

### ARRANGEMENT OF THE STEAM PIPES.

The method given for the arrangement of the hot
water pipes can be followed with few changes for steam,
whether high or low pressure. The main for each house
should be carried along under the ridge to the farther
end, running on a slight decline, where it should be
broken up to supply the coils. If constructed with
manifolds, a manifold valve should be used, or, if in separate lines, all but one pipe on each side should be
arranged so that it can be shut off in mild weather. In
making the coils, and in fact all connections, great care
should be taken to allow for expansion.

For short coils, one-inch pipe may be used, but if of
considerable length, one and one-fourth inch pipe is preferred by most florists. The slope of the coils should be
towards the boiler, when the flow is carried overhead, in

order to return the condensed water. At the end of the house the returns should be collected into one pipe, which should enter the boiler below the water level.

There should be an automatic air valve on each of the coils at the lower end, and on the return, near the boiler, it is well to have both a valve and a check valve.

As recommended for hot water, it is well to have the pipes somewhat distributed, and if, in addition to the overhead mains, one return pipe is carried along on the wall plate, it will tend to warm the cold air that enters through the ventilators, or cracks in the glass, before it comes in contact with the plants. With plants like cucumbers and roses, that are susceptible to cold drafts, this will be found a decided advantage.

## AMOUNT OF PIPE FOR STEAM.

The amount of pipe, both for mains and coils, will be much less than when hot water is used. For the main it can be reckoned that a

1½ inch pipe will supply 200 square feet of radiation.
2    "      "      "      400    "      "      "
2½   "      "      "      800    "      "      "
3    "      "      "    1,600    "      "      "
4    "      "      "    3,200    "      "      "

The surface of the steam pipes is from thirty to fifty per cent. warmer than that of hot water pipes, and a corresponding decrease of the necessary radiating surface can be made. For low pressure steam, in addition to the mains, a house will require for each 1,000 square feet of glass, to warm it to

45 to 50°, 140 square feet, or 300 linear feet, 1¼ inch pipe.
50 to 60°, 175    "      "      "  400    "      "      "      "      "
60 to 70°, 225    "      "      "  500    "      "      "      "      "

With high pressure, a considerable reduction can be made from the above.

In figuring the capacity of a boiler, about fifteen feet of heating (fire) surface should be reckoned as one horse power, and in estimating the radiation that it will supply, from fifty to ninety square feet of radiation per horse power, according to the pressure, may be relied upon with a good boiler.   If we consider that for a temperature of fifty degrees, which may be taken as about an average, one square foot of radiating surface will take care of six square feet of glass, one horse power will be sufficient for 300 to 540 square feet of glass.   As in the case with hot water heaters, a large steam boiler will handle more glass to a square foot of grate than a small one.

The size of grate for a given glass area will also depend upon the draft of the chimney, the skill of the fireman and the method of stoking used.   With a poor draft a much smaller amount of coal can be burned, per square foot of grate, than when the draft is strong, and a grate area considerably larger than in the latter case will be required : the same is true of a dirty fire as compared with a clean one.   For establishments with less than 10,000 to 12,000 square feet of glass, a night fireman can hardly be afforded, and a large grate should be used upon which a slow fire can be burned that will last from six to ten hours.   For this purpose the grate should have an area of from fifteen to eighteen or even twenty feet, according to the climate and other modifying conditions.   On the other hand, when a strong draft can be secured, and in large establishments, where a night fireman is employed, one square foot of grate can readily handle one thousand square feet of glass.   In other words, a steam boiler with twelve square feet of grate can be made to heat with economy 12,000 square feet of glass.   Under favorable conditions, eight square feet of grate will heat a house containing the above amount of glass to fifty degrees.

The matter is so important that it is well to again mention the advisability of putting in a boiler with a capacity twenty-five per cent. larger than is required to do the work, and of arranging for ample radiating surface.

The only other matter of real importance in arranging a system is to have the pipes with such a fall (one inch in twenty feet will answer) that the water of condensation can readily drain off. This can best be secured, if there is a gradual descent in the pipes from

FIG. 74.  INTERIOR OF STEAM-HEATED HOUSE.

the point where the main enters the house to where the return leaves. If it becomes necessary to change the direction of the slope, a one-inch drip pipe should be connected with the underside of the main, at the point where the direction of the pipe changes, and joined to the returns.

ANOTHER METHOD OF PIPING.

Although the overhead main will generally give best satisfaction, particularly in long houses, it is sometimes preferred to have them all under the bench. The coil can commence at the end of the house nearest the boiler, and with a gradual fall to the other end, from which point the return can descend to the boiler. These coils can be underneath the side benches or in the walks, and, if desired, in wide houses under the center bench, also. This method of distributing the pipes will be particularly desirable when the plants are placed over on the benches.

In short houses the coils can run entirely around the house, although the short runs will be preferable. With low pressure it is not advisable to have coils more than two hundred feet in length. Even for houses of this length, it will be very convenient to have the different houses in the range connected in the center by a cross gallery, in which the boilers may be placed, and through which the mains and returns can be run and connected with coils which will be half as long as the house. Fig. 74 shows one method of piping a small house for steam, the furnace-room being at the farther end of the house.

Various methods of arranging steam pipes are shown in Figs. 58-62. As a rule, a two-inch steam main can be used instead of a three-inch hot water main, and a one-inch steam pipe will be equivalent to an inch and one-half hot water pipe in the coils for low pressure, and a two-inch pipe if the steam is under high pressure.

# CHAPTER XX.

The following are among the claims made by advocates of steam for their favorite heating system: (1) A lower first cost; (2) ability to maintain a steady temperature; (3) readiness with which the temperature can be raised or lowered if desired; (4) economy of coal consumption; (5) ease with which repairs can be made.

The hot water men admit that these claims hold to a large extent against hot water in four-inch pipes, but they contend that the men who make these claims have made no comparison with modern well-arranged hot water plants, and that, under proper conditions, the latter system is preferable. Those who favor hot water claim for that method that at the most only the first claim of the steam men will stand, and that on the other points, hot water can make as good, if not better showing.

With regard to the first cost, as stated before, the amount of radiation required for hot water with an open tank is about forty per cent. more than with steam, which will make the cost of the plant about twenty per cent. more than the cost of a steam plant. Under pressure, however, the cost will be little if any more, but we shall lose in economy of fuel, as compared with the open tank system, although it retains all of the other advantages claimed for hot water. With a fireman giving constant attention to the boilers, a steady pressure can be maintained, and of course the pipes being all of the time at the same temperature, there need be but little

9     129

variation in the house, provided the pressure is raised or lowered, or the valves are used to regulate the amount of radiation, according to the outside temperature.

In small plants, where regular firemen are not employed both for night and day, the pressure will vary to a greater or less extent. In well-arranged plants, boilers can be left in severe weather for six or eight hours, and pressure will be maintained, provided everything is all right; but if for any reason the water in the boiler drops below 212, the steam pipes will cool, and serious harm may result. With hot water, circulation will go on so long as there is fire in the heater, and the water in the pipes will give off heat even after that, until they cool to the temperature of the house. It can then be claimed for hot water, and no one can deny it, that in small plants, *hot water is safer than steam to use, and can be left for a longer time without attention.*

It is also urged in favor of steam, that in long runs the hot water becomes cooled, and that the temperature at the lower end of the coils will be less than at the other. In a short house of one hundred and fifty feet or less, this can be counteracted by using the overhead main and underbench returns, and even in long houses the difference, with this method of piping, should not exceed five degrees in a house two hundred feet long. If the continuous coils of one and one-half or two-inch pipe running through the house and back are used, which may be done where a fall of one inch in ten feet can be secured, there may be even less difference than is found in steam pipes.

There is, then, some ground for the claim of hot water men that, even compared with large plants in which night firemen are employed, *the temperature at the opposite ends of the houses will be as even as with steam, and that the hot water system properly arranged will maintain for eight or ten hours a temperature as even as will be secured from steam by the average fireman.*

The claim that steam can be used to better advantage when it is desired to raise or lower the temperature of the house, only applied against water in large pipes. If desired, the entire circulation in the hot water coils can be shut off, and the amount of heat in the water in the pipes if given off at once would not raise the temperature of the house a single degree, and distributed over an hour or so, would not be noticed. With four-inch pipes containing ten times the quantity of water, and, especially as valves were not always provided for shutting off the circulation, the heat given off was sufficient to necessitate the early opening of the ventilators on bright mornings and a corresponding injury from cold drafts upon the plants was caused. With small pipes, starting with cold water or with a moderately low fire, a normal temperature of the pipes can be secured as quickly as with steam. When the question of economy of fuel is considered, the general opinion of those who have carefully tested steam against the modern hot water system, is that the latter is about twenty-five per cent. cheaper.

## EXPERIMENTAL TESTS.

There are on record a large number of so-called tests of the economy of steam and hot water, but in nearly every case the hot water was in four-inch pipes. The only experimental tests that have come to the notice of the writer, where the houses and plants were of similar construction, and the tests were carried on at the same time, was the one by Prof. Maynard, at the Massachusetts Experiment Station at Amherst, and those of the author at the Michigan Experiment Station. In each case piping was arranged in both houses with overhead mains and underbench coils, and although an attempt was made to have each plant as perfect as possible, the conditions for either system were no more favorable than could be secured in any forcing house. The houses at Amherst were seventy-five by eighteen feet, and those at Lansing twenty by fifty feet each.

The tests were continued in each place for two years with the following results:

| | AVERAGE TEMPERATURE | | COAL CONSUMED. | |
|---|---|---|---|---|
| | Water House | Steam House | Water House | Steam House |
| Amherst | 52.50 | 50.80 | 69.54 lbs. | 91.48 lbs. |
| Lansing, | 4.57 | 5.02 | 94.53 " | 114.53 " |
| Average | 3.54 | 4.91 | 82.04 " | 103.01 " |

Or almost exactly twenty-five per cent. more fuel was required for steam than with water, although the steam houses averaged about two degrees cooler.

In both places the hot water house was more exposed to the cold winds than the steam house, and, at Lansing, where the results were less favorable for water than at Amherst, although the houses were piped for maintaining a temperature of forty-five to fifty degrees, they were kept at a temperature of fifty-five degrees, necessitating a considerably higher temperature of the water than should have been carried for the greatest economy of fuel, which would make less difference with the steam system. In proof of this, additional pipes have now been put in, and the hot water house is now carried at sixty degrees with no more fuel than was used at fifty-five degrees.

### COMPARATIVE COST OF FUEL.

With steam it is claimed that a cheaper grade of fuel can be used than with hot water, but boilers for hot water are now made that can secure better results from soft coal than is obtained by steam, provided similar care is given. It is also claimed for steam that it admits of the boilers being located in a battery at one point, rather than scattered in different houses as is generally the case with hot water. With modern systems of piping, the same arrangement can be used, and the water can be carried in large mains with less waste than when steam is used.

Regarding the last claim in favor of steam, *i. e.*, the economy of repairs, it may be said that when the same size pipe is used for coils in both systems, a break can be repaired with the same ease in one as in the other. Moreover, there is more likely to be a break in the steam boiler than in the hot water heater, and while a steam return is rusted through in from five to seven years, a hot water pipe will be found in good condition *inside*, and the outside can be kept from rusting by painting once in two or three years with lampblack and oil.

## CONCLUSIONS.

Considered from the point of efficiency only, there is little to choose between the systems, although the steam heater will need more constant attention, and ordinarily the temperature of the houses will be less regular than with hot water, either with open tank or under pressure.

The steam plant will cost fifteen to twenty per cent. less than the open tank water system, and somewhat less than the pressure system, and when the first cost of the plant is any object, this may decide for that system. On the other hand, the cost of fuel with a well-arranged hot water plant, will be twenty to twenty-five per cent. less than with steam, and as this will pay for the extra cost of the plant in three or four years, it becomes a matter well worth considering. It then comes to the question whether there shall be a large cost at first, with a comparatively small outlay for fuel and repairs, or a smaller first cost, and a larger outlay for fuel and maintenance.

Everything considered, the man who has less than 10,000 square feet of glass, will find hot water with an open tank the best method to use. Above 12,000 feet of glass, it will pay to have a night fireman, and, as the first cost for a plant of this size is considerable, the average florist will prefer to use steam, although hot water will give fully as good results, and the extra expense of the plant will be saved in fuel within four years. So far as expense for fuel is concerned, hot water under pressure will be classed with steam; it gives more even results, however, and the cost of the system is little if any more. In arriving at these conclusions, no account is taken of the effect of the different systems upon plant growth, as we believe that when equally well cared for there will be little or no difference.

## CHAPTER XXI.

### HEATING SMALL CONSERVATORIES.

For amateur conservatories, with over 300 square feet of glass, unless joined to a residence which is heated by hot water or steam, it will be found desirable to use some of the small portable hot water heaters that are manufactured by several firms. When these are used in connection with a well arranged system of piping, the care of the house is greatly simplified, and there will be little risk of injury to the plants by cold. It will be a desirable thing, if the dwelling is heated with stoves or a hot air furnace, to purchase a heater large enough to warm a part or all of the house, and put in pipes and radiators.

In arranging the heating system for the conservatory, the heater should be placed in the cellar of the house, and the feed pipes should pass up through the floor and connect with the radiating pipes, which are generally best if arranged in a wall coil, with manifolds at each end. An air valve will be needed at the higher end, and an expansion tank should be connected with some part of the system. It should be of galvanized iron, although an old paint keg would answer. It should hold a gallon for each hundred feet of one and one-fourth inch pipe, and a gallon for the heater and mains. If the tank is situated where harm to floor or walls will be done if it boils over, it is well to have a tight cover on the tank and run an overflow pipe from half-way up the side of the tank to a drain.

For conservatories which are too large to heat with an oil stove, a home-made water heater might be used. The radiating coil and attachments would be similar to those just described. A small heater could be made by using a small coil of one-inch pipe containing eight linear feet of heating surface, inside one of the large sized kerosene heating stoves. This would warm one hundred and fifty linear feet of one-inch pipe, and would heat a conservatory six by ten feet with three sides and roof of glass. Were the conservatory upon a veranda where only the roof, side and one end were exposed, the capacity would be sufficient to warm about six by fifteen feet, and if the roof were of wood, it could heat a space eight by twenty feet.

### THE BARNARD HEATER.

In 1890 Charles Barnard described in the *American Garden* a very simple heater that gave good satisfaction in a detached greenhouse. The heater was of zinc with four tubes of one-inch gas pipe (Fig. 75 *A*); the diameter was six inches and the height twenty-seven inches. From this, connections were made with the coils which were of two-inch pipe, although one and one-quarter inch pipe would be preferable. The heater was placed over an oil stove or gas burner, and was surrounded by a jacket of sheet iron (Fig. 75 *B*) from which a small pipe ran to the outside of the house to convey the smoke and gases. Another form of heater is made in the shape of a hollow truncated cone, nine inches in diameter at the bottom and six at the top, and twenty-four inches high. The water is one inch in thickness and is confined between the inner and outer shells of the heater. This is placed over an oil stove and arranged in much the same way as the one last described.

## HEATING BY MEANS OF FLUES.

In small houses where one does not have the means to put in hot water or steam, fairly good results can be

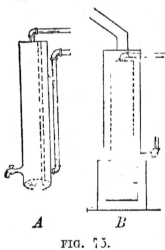

obtained with the old-fashioned flue. This consists of a furnace in which the fuel is burned, and a horizontal chimney passing through the house. If the house is not over fifty feet in length, and if a rise of two or more feet can be secured, a fair draft can be obtained by having the chimney at the farther end; but in longer houses, or where the flues must be run on a level, it is best to bring them back, so that they can enter a chimney built over the furnace.

*A*      *B*

FIG. 75.

BARNARD HEATER.

A direct connection with the chimney can be made when the fire is first started, and then, after the chimney has become warm, a damper can be turned which will force the smoke to pass around through the house, giving off its heat as it goes. The furnace can be constructed for burning either coal, or wood cut in lengths of from three to five feet. A grate containing three to four square feet will answer for a house containing 600 square

feet of glass. If wood is used, the furnace should be
eighteen inches wide inside, and of the required length,
but no increase of the size of the grate will be necessary.
There should be an ash pit of suitable size, and iron
doors should be set in the masonry at the end of the
furnace, for both the fire-pot and ash pit. The top of
the furnace may be supported either by a brick arch or
by heavy iron bars. The inner lining of the heater
should be of fire brick laid in fire clay, and the same
material should be used for the first fifteen feet of
the flue. Beyond this point, common stock brick will
answer, forming a flue eight by twelve to sixteen inches,
or eight to ten-inch glazed tile may be used.

For a house twelve feet in width, one flue will
answer; but if fifteen to twenty feet wide, it is well
either to have a return flue on the other side, or to
divide the flue and carry up a branch on each side,
either under the walks or beneath the side benches.

A hot water coil can be economically combined with
a flue by using cross-pieces of one and one-half inch
pipe, connected by return bends, across the side walls and
supporting the top of the heater, and connecting them
with the radiating pipes. If a flue is used care should
be taken that no woodwork comes in contact with the
bricks within thirty feet of the furnace. When houses
are very long, furnaces may be placed at both ends and
the flues can be carried half the length of the house and
brought back on the opposite side.

### THE POLMAISE SYSTEM.

The Polmaise system was so-called from the French
town where it was first used. The original system consisted in bringing a current of air over a heated surface,
and then carrying it into the greenhouse, on its way
passing through a wet blanket, that its drying effect
might be lessened. The system itself is of no value, but

a modified form of it may be used in connection with a flue. By building an air chamber around the furnace and admitting the air, much as in common hot air furnaces, it will be warmed, and can be carried through the house in *tiles* much as are the products of combustion.

The cost of a flue is less than half that of a hot water or steam plant, and especially if combined with hot water, as described, very satisfactory results can be obtained. The modified Polmaise system could also be employed with profit, if the coil is not used.

### FIRE HOTBEDS.

In addition to warming hotbeds by means of decomposing manure, various other methods of heating have been tried, the simplest being a modified form of the ordinary flue as just described. The beds can be single, for sash six feet long, or can be double span-roofed structures, with a row of sash on each side. For the single beds the arch or furnace need not be over one foot wide inside, eighteen inches high, and four or five feet long. It should be arched over with brick and the whole then covered with soil. In order to secure proper slope for the flues, the hotbeds should be located on a hillside sloping to the south, and the flues should have a slope of about one foot in twenty, although more is desirable. The tile used for the flues should be glazed for the first twenty feet at least and six inches in diameter, and should be laid in two lines, three feet apart; at the farther end the tile should be turned up at right-angles forming a chimney. An ordinary hotbed frame should be set over this. The soil at the furnace end should then be spread on, covering the arch to the depth of twelve to fifteen inches, and the pipes at the chimney end about six inches. The draft can be regulated by a plate of iron resting against the end of the arch. The structure will last several years, and will prove a great convenience where one does not have a greenhouse in which to start vegetable plants, and where wood is cheap.

For the span-roof hotbed, two arches or furnaces and four flues, arranged as in the other case, will be required.

### STEAM AND HOT WATER HEAT FOR HOTBEDS.

If it is desired to warm hotbeds by means of steam, it can be done by running a one and one-quarter inch steam pipe up in one line of four-inch drain tile, and back in another line laid as described for the flues with the narrow beds, while four lines would be required for a bed twelve feet wide. When exhaust steam is at hand it can be used without the steam pipe by merely discharging it into the tile.

A frame can be heated by hot water or steam if a two-inch hot water or an inch and a quarter steam pipe is run around the inside, next to the plank. Boards should then be placed so as to shut off all direct heat from the plants. If a crack two inches wide is left between the top of the boards and the glass, the heat will be diffused and will not dry out the plants.

## CHAPTER XXII.

### COMMERCIAL ESTABLISHMENTS.

A florist just starting in business may be compelled by lack of means to commence upon a small scale. While he would find a lean-to house the cheapest to erect—provided he built it against the south wall of a building—the excess of cost for a span-roof house would be so slight, and the results obtained would be so much greater, that he would be wise in selecting that form for a house. The size for the house must be determined by the business to be done, but for most purposes a house of twenty feet in width is preferable to anything narrower, and an enterprising florist should be able to utilize one that is fifty feet long.   It is desirable to have both

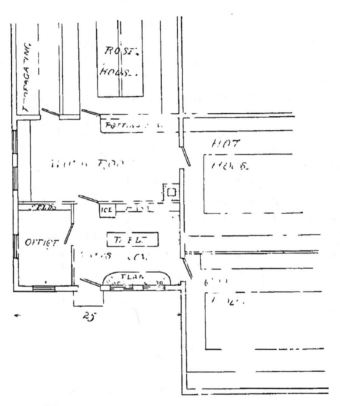

FIG. 76.   PLAN FOR A SMALL ESTABLISHMENT.

a cool and a warm house, and this can be secured by running a glass partition across the house.

If this amount of glass is not sufficient, a second house can be built similar to the first one, and then he will have one house to be kept at a temperature of fifty-five to sixty degrees and another that can be kept at forty-five to fifty degrees.   Although other houses are desirable, a good selection of plants can be grown in two such houses with fair success. If business develops, as it should, it will be desirable to add a rose house. This should be of the three-quarter span form, eighteen and one-half feet wide, and will give an opportunity for the erection of a north side propagating house, which can not only be used for propagating, but will be excellent for ferns, violets, pansies, and for the starting of seeds and bulbs. The even span houses could run north and south with a workroom at the north end twenty-five by twenty-five feet, while the rose house could join the end of the workroom and run east and west, as shown in Fig. 76. A convenient arrangement for the workroom and store is shown in the illustration, which can readily be understood.

If still other enlargement of the establishment becomes necessary, the additional buildings can be put up parallel to the present ones, or they can be run out the other way from the workroom. Another method would be by lengthening the buildings already put up, but for small establishments it will hardly be desirable to extend them beyond a length of one hundred and fifty feet.

In addition to the general· florist and vegetable grower, we find to-day engaged in greenhouse work many specialists, and among these the commercial rose grower and the lettuce grower, from the extent of their business, are especially worthy of notice. As in everything else, we find, as a rule, that these specialists who have turned their every effort to the doing of one thing well, are masters of their business, and have been quick to avail themselves of all the latest improvements.

# CHAPTER XXIII.

The form and general arrangement of the houses used for forcing roses, is practically the same the country over, and when one speaks of a "rose house," he is readily understood. A rose house may be briefly defined as a three-quarter span greenhouse, about eighteen feet wide, with two narrow beds at the sides, and with two somewhat wider ones in the center. No form of house has been tried for this purpose that is on the whole as satisfactory as this, of which a good example of an exterior will be found in Fig. 77.

They are cheapest to build and easiest to heat if constructed with wooden walls up to the plate, as shown in Fig. 82, but many of our best rose growers are of the opinion that the extra cost of erection and maintenance is more than repaid by the results obtained, when there is from eighteen to twenty-four inches of glass in the south wall and ends under the plate. There seems to be a diversity of opinion as to the best width for rose houses, the range being from sixteen to twenty feet ; but it is the general idea that in the houses sixteen feet wide there is a lack of economy of space, unless the walks are made rather narrow. With the side walks eighteen to twenty inches wide, and a walk between the center benches with a width of twelve inches, there will be room for four benches of average widths ; but for convenience the walks at the side should not be less than two feet in width, and the center walk from fifteen to eighteen inches. A convenient width for the front bench is thirty inches, which will answer for three rows of plants; the center beds should be three feet and six inches, each holding four rows, and the back bed two feet in width with two rows of plants. If the front wall is made six inches, and the rear one eight inches in thickness, with the benches set out to prevent drip from the plate, a house with the above widths for walks and benches will be about eighteen feet and six inches to the outside of the walls. In locating the height of the benches, the tops of the cross bearers for the front and back bench should be about twenty inches below the plates; the south center bench should be at the same height as the front bench, and the north one about eighteen inches higher. Some growers prefer to have both of the center benches level, but if careful attention is given to the watering, rather better results will be obtained if they are given a slight slope to the south, say of eight inches in the width of one bench or of eighteen inches between the walks (Fig. 63).

It is quite desirable in arranging the roof to have the ridge and purlin come over the walks. If an iron frame-work is used with a truss at the ridge, there will be no necessity for a support under the ridge; but if the roof is of wood, particularly if there are no principal rafters, a post should be used, and the ridge should be so located that the post can pass down at the north side of the center bench. While one purlin with one row of posts—in addition to the one under the ridge—will support a roof of this width, lighter material can be employed, and there will be less trouble from drip if two of each are used, with the posts coming down at the south side of each of the center benches.

Particularly in rose forcing houses, it is desirable to have the slope of the roof arranged to trap as much as possible of light and heat from the sun during the winter months, and, everything else considered, the south

FIG. 77. W. P. WIGHT'S ROSE HOUSES, ERECTED BY THOS. W. WEATHERED'S SONS.

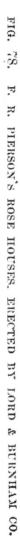

FIG. 78. F. R. PIERSON'S ROSE HOUSES, ERECTED BY LORD & BURNHAM CO.

10

pitch of the roof should slope at the rate of about two
feet for every three feet in width of house. With the
ridge posts at a distance of fourteen feet from the outside of the south wall, the bottom of the ridge should
be about eight feet higher than the top of the south wall,
or twelve feet from the ground level, with the south wall
four feet in height. This will require a rafter slightly
less than sixteen feet in length on the front, and six feet
on the rear slope of the roof, when the rear wall is eight
feet in height. Another good form for a commercial
rose house is the one described in Chapter III., with the
sides of the roof fifteen and seven and one-half feet, and
the height of the front and back walls five and seven
feet respectively. In a house of this shape there should
be a line of glass under the plate of the south wall (Fig.
77). While the even-span house is not as well adapted
for rose forcing as the three-quarter span house, it is frequently used, and will give very fair results. These
houses may be eighteen to twenty feet wide, with four
benches, about three and one-half feet each, in width.

The best results seem to be obtained from benches
not over four inches in depth, although this varies with
the character of the soil, as three and one-half inches of
heavy soil will be equal to four and one-half inches of
soil of a sandy nature. In selecting the material for the
bottoms of rose benches, a first choice would be for tile,
second slate, and third wood. In planning our rose
houses everything has been arranged upon the presump-
tion that shallow beds were to be used, as this seems to
be the favorite method of growing them.

When there is no glass beneath the plate on the
south wall, the custom in the past has been to have a
single line of ventilators at the ridge, but many of the
more recently constructed houses have a line of sash on
each side of the ridge; if these are properly used, the
draft of air upon the plants is greatly decreased. The

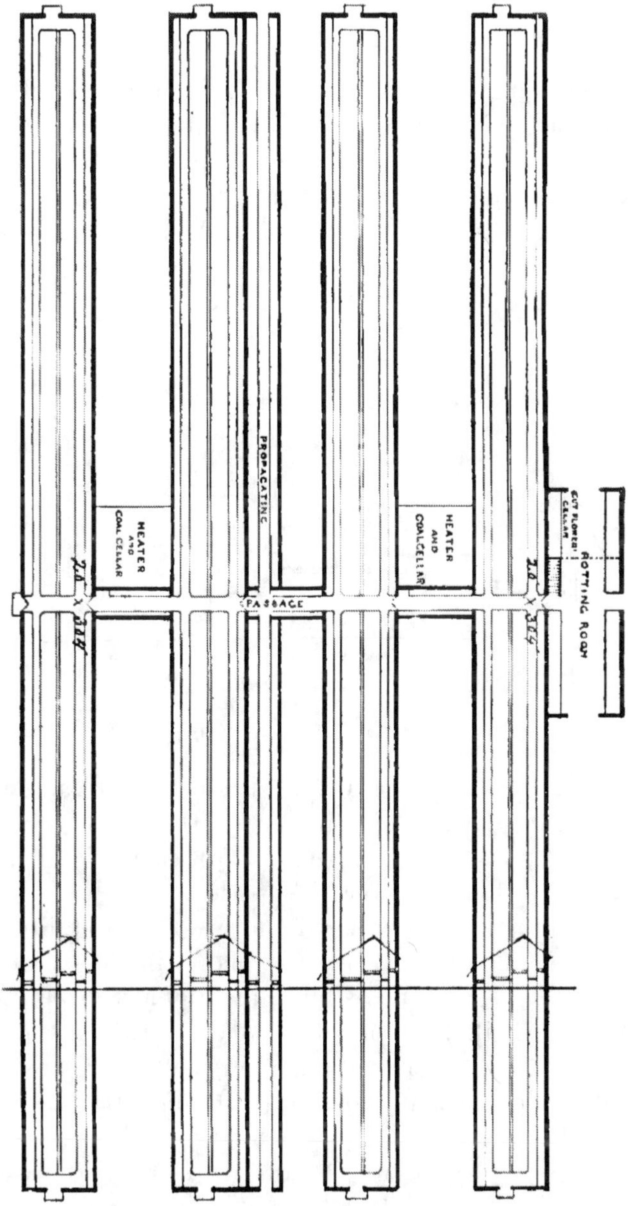

FIG. 79. GROUND PLAN OF F. R. PIERSON'S RANGE.

use of ventilating sash in the south wall is also quite
common.

During the last five years many large and well
arranged commercial rose houses have been erected, and
we are glad to be able to show illustrations of two plants
that contain many of the latest ideas. In Fig. 77 will
be seen a perspective view of the rose houses erected for
W. P. Wight of Madison, N. J., by Thos. W. Weath-

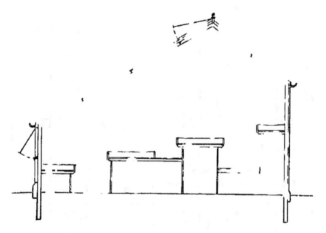

FIG. 80.   SECTION OF IRON ROSE H...

ered's Sons of New York City. From this we can get an
idea of the general appearance of the better class of
commercial rose houses. They are each about three
hundred feet long by twenty feet wide.

One of the best arranged and most thoroughly constructed commercial greenhouse plants in the country is
shown in Fig. 78. It was built in 1890 at Scarborough,
N. Y., for F. R. Emerson, by Lord & Burnham Co. As
will be seen by the ground plan Fig. 79, there are eight
houses each measuring one hundred and fifty by twenty
feet. They are placed in pairs, end to end, except for a
narrow passage way which affords a ready means of
communication with the different houses and with the

potting shed. At the rear of the second line of houses is a propagating house, nine feet wide and three hundred feet long, or really two houses each one hundred and fifty feet. The construction is the same as is recommended in Chapter VII., the rafters and posts being of iron, with the lower ends of the latter set in cement in the ground. The purlins and ridge are also of iron and all woodwork of cypress. The benches have an iron frame and slate bottom. The heat is furnished from steam boilers located as shown in the ground plan. A cross section of one of these houses is shown in Fig. 80, from which one can obtain a good idea of the slope of the roof and of the interior arrangement, while Fig. 81 shows another method of construction.

A house of this description can be erected for about $25.50 per running foot, including the steam-heating apparatus. With hemlock benches the cost would not be over $20.75, and were the glass left out under the south plate, leaving only one line of ventilating apparatus at the ridge, the cost could be reduced to $20.00 per linear foot. This would give double strength, French "seconds" or American "firsts" glass, and two coats of paint. Reckoning the steam-heating apparatus at $4.50 per linear foot, the house complete as above, with cheap wooden benches and without heating apparatus, would cost something over $15.00 per foot. When iron benches with wooden bottoms are used, the house with one row of ventilating apparatus and steam-heating apparatus, would cost not far from $22.00 per foot. If one does not care to use the iron construction, cypress lumber can be obtained for the erection of a rose, or, in fact, of any kind of a greenhouse, all gotten out in the most approved sizes and shapes, ready to be fitted together.

There are a half-dozen or more firms who make a specialty of cypress for greenhouse building, among the oldest of which is the Lockland Lumber Co., of Lock-

FIG. 81. INTERIOR OF ROSE HOUSE, ERECTED BY PLENTY HORTICULTURAL WORKS.

land, Ohio. By request they have prepared a ground
plan, cross section and details of a rose house such as
they furnish, and they are here presented as suggestions
to prospective builders. The cross section is shown in
Fig. 82 and is so clear that any carpenter could put the
house together. The *details* are shown in Chapter VI.
and afford us an idea of some of the best shapes for the
different parts of a greenhouse, and the way to put them
together. The patterns do not differ materially from
those used by other dealers in greenhouse materials, and

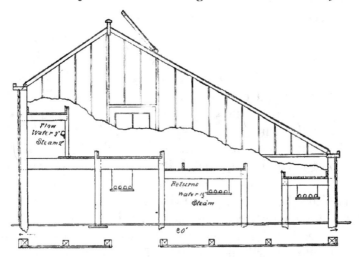

FIG. 82. SECTION OF ROSE HOUSE (WOOD).

perhaps the best advice that could be given to a person
intending to build a rose house would be, in case one
could not afford to build a house with an iron framework, to write to the nearest dealer in cypress lumber for
plans and estimates for the proposed structure.

As a partial guide in the matter, the following estimate is offered as the probable cost of cypress lumber
for the erection of a three-quarter span rose forcing
house. This includes all lumber required above the
walls for a house with one glass gable; the lumber being

dressed upon four faces and worked to proper shape with dimensions as given, including door and one row of ventilators.

ESTIMATE FOR CYPRESS LUMBER FOR A THREE-QUARTER SPAN FORCING HOUSE, 100 FEET BY 20 FEET.

| Gable Plate, | $1\frac{3}{4}'' \times 7''$ | | 20′ long | @ | $0.07\frac{1}{2}$ | $1.50 |
| Side Gutters, | Bottoms and two sides, | | 200′ | @ | .11$\frac{1}{2}$ | 23.00 |
| Ridge Pieces, | $7\frac{1}{2}'' \times 1\frac{3}{4}''$ | | 100′ | @ | .08 | 8.00 |
| Ridge Cap, | $1\frac{1}{4}'' \times 3\frac{1}{2}''$ | | 100′ | @ | .02$\frac{1}{2}$ | 2.50 |
| Purlin, | $1\frac{3}{4}'' \times 3\frac{1}{2}''$ | | 100′ | @ | .03$\frac{1}{2}$ | 3.50 |
| End Rafters, | $1\frac{3}{4}'' \times 3''$ | 2 of 7′ 2 of 16′ | 46′ | @ | .03$\frac{1}{4}$ | 1.61 |
| Gable Sash Bars, | For one gable, | | 80′ | @ | .01$\frac{3}{4}$ | 1.40 |
| Roof Sash Bars, | | 82 of 7′ 82 of 16′ | 1,880′ | @ | .02$\frac{1}{4}$ | .43 |
| Header, | $1\frac{1}{4}'' \times 2\frac{1}{2}''$ | | 100′ | @ | .02$\frac{1}{2}$ | 2.50 |
| Ventilators, | 3′ wide 1$\frac{1}{4}''$ | | 100′ | @ | .16 | 16.00 |
| Door and Frame, | Door 3′ x 7′ x 1$\frac{3}{4}''$ | | | | | 4.75 |

$107.19

For the construction of the walls twenty-six posts seven feet long, and costing about twelve cents each for cedar, and twenty-one posts twelve feet long, which will cost about twenty-five cents each, or about ten dollars for posts, will be required. Red cedar will cost two or three times as much, and locust, which will be found very durable, will vary in price but will generally cost less than red cedar. For sheathing the building 1,300 feet of matched hemlock, costing from $10.00 to $15.00 per thousand, will be required, and the outside siding will take 1,500 feet, which will cost about $20.00 per thousand. A small amount of finishing lumber, building paper and nails will complete the exterior, with the exception of the painting and glazing. The interior will require tables and walks, gas-pipe posts and ventilating and heating apparatus, which, with hinges and other hardware for door and ventilators, will cover the necessary materials for the erection of a forcing house.

The cost of the lumber will be about $230.00; glass, 2,500 feet at $3.50 per box for double strength B, $145.00; ventilating apparatus, $30.00; nails and hardware, $6.00; paint and putty, $50.00; building paper, $5.00; gas-pipe posts, $12.50; making a total cost of materials for the house, exclusive of labor, of about $500.00 to $525.00. The heating apparatus, water supply, etc., will be additional. The former will cost about $400.00 if hot water is used, and not far from $340.00 for steam, including labor. If the house is erected by hired labor the cost will be from $250.00 to $300.00 for the carpenters and painters, according to the experience of the men and the wages paid.

Briefly summarized then, the cost of a three-quarter span forcing house complete with heating apparatus will be:

| | |
|---|---:|
| Lumber for walls | $60.00 |
| Lumber for roof and freight | 115.00 |
| Lumber for benches | 45.00 |
| Lumber for walks | 10.00 |
| Glass, 50 boxes 16″ x 20″ | 175.00 |
| Paint and putty | 50.00 |
| Ventilating apparatus | 30.00 |
| Hardware | 6.00 |
| Building paper | 5.00 |
| Gas-pipe posts, 1¼ inch and 1 inch | 12.50 |
| Labor, carpenters | $125.00 to 150.00 |
| Painters and glaziers | 125.00 to 150.00 |
| Heater, smoke pipe, etc. | 200.00 |
| Pipe, valves, and fittings | 100.00 to 150.00 |
| Labor | 40.00 to 50.00 |
| Total | $1,098.50 to $1,208.50 |

Or, $11.00 to $12.25 per linear foot.

In the above estimate, the grading, drainage, water supply, and the cost of a potting shed, furnace cellar, etc., are not considered. Of course, the cost of the lumber and the expense for labor, etc., would vary in different localities, so that no estimate can be made that will apply in all cases, but where lumber can be obtained at from $12.00 to $20.00 per thousand, according to the grade, and labor of carpenters and painters is not over $2.50 per day, the above will be sufficiently reliable to furnish a fair idea of the cost. The cost of an even-

span house will be about the same, and, if the back wall
has to be built for a lean-to, it will cost fully as much as
the others for the same width of house. In this estimate over $300.00 is allowed for labor, and as many
florists would do most of the work themselves, a considerable reduction could be made in this item.

## CHAPTER XXIV.

### LETTUCE HOUSES.

Although we still find many growers of lettuce using
houses of lean-to or narrow even span construction, the
wide houses are rapidly superseding them. Perhaps
the largest house ever erected for the purpose was constructed by W. W. Rawson of Arlington, Mass., who
has been engaged in the growing of lettuce and other
garden produce for the Boston market for many years.

This house is three hundred and seventy feet long
and thirty-three feet wide. It is of the three-quarter
span form, and measures fifteen feet high at the ridge,
with a south wall three and one-half feet high, and the
north one twelve feet in height. The glass is double
strength, twenty by thirty inches. The crop from this
one house is about two thousand dozen heads, which
sometimes brings from $2,000.00 to $2,500.00. Three
crops are grown in a year besides a crop of cucumbers.
While this is the largest house of the kind, there are
many smaller ones constructed upon the same general
lines, and they seem to be uniformly successful.

### LEAN-TO LETTUCE HOUSES.

The lettuce is a plant that succeeds well in a lean-to
lettuce house, such as is used by many of the lettuce

growers in the vicinity of Boston, of which a cross section is shown in Fig. 83. Like all lean-to houses these are easily warmed and are cheaply constructed, but they do not have a sufficient pitch to the roof to secure the most benefit from the sun. They are commonly given a pitch of about eighteen degrees, but even at this slope, a lean-to roof on a house thirty-three feet wide would require a north wall about fifteen feet high, while a three-quarter span house can have a pitch of twenty-two degrees, and the north wall need not be over ten or twelve feet high.

With houses up to a width of twenty-five feet, a proper slope can be secured without carrying the north

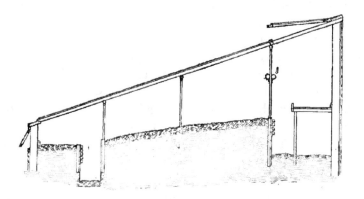

FIG. 83. LEAN-TO LETTUCE HOUSE (*Section*).

wall to an undue height, or raising the glass too high above the plants; unless upon a sidehill, this width cannot be very much exceeded with this style of house. The three-quarter span house can readily be made eight or ten feet wider than the lean-to, without carrying the roof to a greater height, while the north wall will be considerably lower than it would be in the lean-to. Having determined upon the width and style of roof for the house, the construction will be very simple, if the sug-

gestions given in Chapters V. and VI. as to the best
methods of erecting the walls and roof are followed.

Lettuce houses should have from eighteen to thirty
inches of glass in the south wall, and, on many accounts,
it is desirable that the alternate sash at least be arranged
as ventilators.  Particularly in the wide houses with flat
roofs, the sash bars should be somewhat heavier than for
small houses with steep slopes, and should be very carefully supported.  For lean-to houses twenty-five feet wide,
there should be at least three rows of purlins and purlin
posts.

A house of this description will require a back wall
at least ten feet high, if built on level ground.  A post
and double-boarded wall will be fully as satisfactory as
one built of brick or masonry of any kind.  For convenience in handling the soil, and to assist ventilation,
it is well to have small windows, perhaps two and one-
half feet square, once in ten feet, in this wall.  When
solid beds are used, the south wall should not be more
than three and one-half feet high, although it may be
somewhat higher if the lettuce is grown on raised beds.

The side-hill houses (Fig. 8) will also be found quite
desirable for lettuce forcing, as they are nothing more
than a number of lean-to houses placed close together,
and they will be found not only economical in construction and heating, but in land and in labor of handling
the crops, although the three-quarter span house is
generally preferred to either of the above styles.

# CHAPTER XXV.

## PROPAGATING HOUSE.

In connection with every greenhouse there should be a bench for the rooting of cuttings, and in large establishments one or more houses will be required for this purpose. The simplest method of erecting a propagating house, when one has a rose or other three-quarter span house, is shown in Fig. 61. The structure is known as a "north-side house," and, if it is not entirely needed for propagating purposes, can be utilized for ferns, violets, and other plants that thrive without direct sunlight. In arranging some establishments, a narrow house, connecting the ends of the main houses, is often a convenience. If upon the north, or even on the east or west end, as it generally is, this head house can be made so constructed with a lean-to roof, and will serve an excellent purpose as a propagating house. When any of these benches cannot be utilized, a narrow even-span house can be built for this purpose, and will be well adapted for it.

The construction of the house will not differ from that of a similar house for other purposes, but its interior arrangement should be somewhat different. As a table for a propagating house, an ordinary greenhouse bench will answer, but in order to secure and control the necessary bottom heat, the front of the bench should be boarded up, with one board on hinges so that it can be used to regulate the temperature. For the propagation of most plants this bench will answer as well as a

157

more elaborate construction. The heating pipes should
all be under the bench, and should give a radiating surface about twenty-five per cent. greater than would be
required in a growing house for the same plants as are
to be propagated.

## WATER BENCH.

The use of the old-fashioned wooden water bench
has been abandoned, although the galvanized iron water
bench is quite common. If this is used the trough
should be about four inches deep, and of the width and
length of the proposed cutting bench. It should be well
supported, so that there will be no danger of its settling
at any point. The bottom of the cutting bench, upon
which the sand is to be placed, should be just above the
tank. The tank should be connected at each end by
means of one and one-fourth inch pipes with a hot water
heating apparatus. When the heating pipes in the system are all below the level of the tank, no cover will be
required, but, if at any point the piping is overhead, a
closed tank will be necessary, or an independent heater
can be used for the propagating house, in which case
the water tank will answer as the expansion tank for the
system. When tanks are used, the heating pipes should
be sufficient to maintain a temperature of forty-five to
fifty degrees without the tank.

## PROPAGATING CASE.

While cutting of most plant require thorough ventilation many of the species can only be struck with
success in a close, moist atmosphere. When only a few
are to be rooted, the required conditions can be secured
by the use of a bell glass or hand glass, but if many
are to be struck, a propagating case will be a necessity.
This can readily be constructed upon the bench, preferably at the warmer end. The ends and front can be

made of sash bars and glass. A portion of the front,
however, should be of glass sash, arranged to slide by one
another and give ready access to all parts of the case.

## CHAPTER XXVI.

### HOTBEDS.

Among the movable plant structures, we have what
are known as hotbeds and cold frames. They differ
only in the degree of heat they receive, the cold frame
being without artificial heat, while the hotbed is heated
by fermenting vegetable substances, generally stable manure, leaves, and other refuse. The hotbed, in some of
its forms, is a very de-
sirable, and, in fact,
almost a necessary ad-
junct to the green-
house, for all florists
and market gardeners.
On the other hand,
while a large business
can be carried on with

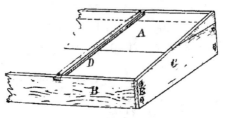

FIG. 84. HOTBED FRAME.

hotbeds alone, the possession of a greenhouse, small and
cheaply constructed though it be, will be a great convenience, particularly to the market gardener, for the
starting of young plants during the severe weather of
midwinter.

The simplest kind of a hotbed, and the one generally used, is about six feet wide, and of any desired
length, with the sash sloping toward the south. While
hotbeds are often made of one-inch boards, or are
cheaply constructed of waste pieces of lumber, they will
be more satisfactory if constructed of lumber that is one

and one-half or two inches thick, carefully framed
together and painted. A very satisfactory hotbed can
be made from three pieces of two-inch hemlock lumber
one foot wide and twelve feet long. In order to give
the sash a proper pitch to the south, one side of the bed
should be made six inches wider than the other. When
planks with a width of twelve inches are used this can
be readily secured, by sawing a strip three inches wide
from the edge of one, and nailing it to the edge of
another, Fig. 84. In this way we secure a plank nine
inches wide ($B$) for the south side of the bed, while that
for the north side ($A$) will have a width of fifteen inches.
The ends should be cut six feet long, and the proper
slope can be given them by sawing off a triangular strip
from one end, and nailing it upon the other end of the
piece, as at $C$, in Fig. 84.

### PORTABLE FRAMES.

A portable hotbed frame is economical, and as it can
be taken apart and stored out of the sun and rain for

FIG. 84. FRAME AND SASH.

six months of the year, it will last for many seasons.
By fastening irons of proper shape to the ends of
each of the side boards, and boring holes to correspond
in the end pieces, the frame can be held together by
washers and pins, as shown at $I$, Fig. 84. Cross bearers, $D$, four inches wide and one inch thick, dovetailed
into the edges of the front and back boards, will keep
the bed from spreading, and will serve as slides for the

sash. If a strip of one-inch board is fastened to the middle of the upper side of each cross bearer it will serve to strengthen it, and to hold the sash in place. Another method of ventilating the bed is illustrated in Fig. 85. For the purpose of retaining the heat in cold weather, straw mats and wooden shutters are desirable.

### MATS AND SHUTTERS.

For the mats, a supply of long rye straw, tarred rope, and strong linen twine are necessary. There are various ways of making the mats, one of the simplest being upon a frame of two by four inch lumber of the same size as the mats. With long straw, a mat six and one-half feet square can be made, but the usual size is

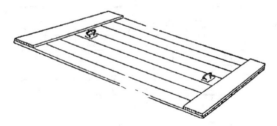

FIG. 86. HOTBED SHUTTER.

about four, or five, by seven feet. The tarred rope is stretched lengthwise of the frame, so as to bring the strands one foot apart and six inches from each side, and fastened to stout pegs. For a mat six and one-half feet square, the straw should be, at least, four feet long. Bundles of straw as large as can be enclosed by the thumb and middle finger, are placed on the frame, with the butts even with the sides, and are tied in place with stout hemp twine. The bundles thus lap in the center about two feet, and the ends will keep the center even with the sides. With straw still longer than this, a mat about five feet wide can be made without any laps of the straw, by placing the butts alternately to the right and

11

left, one length of straw reaching across the mat. If
the mats are kept covered with the shutters, and are
stored where the mice cannot destroy them, they can be
used for many years.

The shutters, Fig. 86, for covering the mats should
be six and one-half feet long, and three feet to three feet
and six inches wide. Made of half-inch matched lumber, with cleats at the ends and across the middle, and
with handles, they form a useful addition to one's
equipment.

### HOTBED YARD.

Unless they can be placed where they will be sheltered by buildings, a tight board fence upon the north,

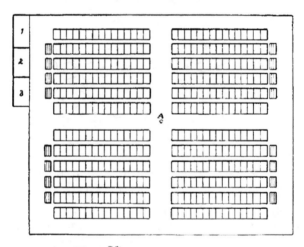

FIG. 87.   HOTBED YARD.

east and west sides will be desirable. The land should,
if possible, slope slightly to the south, and the rows of
frames should be regularly arranged. In Fig. 87 will be
seen a convenient arrangement for a frame yard. There
is an opportunity for a team to pass entirely around the
frames, and through the center in either direction.
There should be a hydrant for furnishing water at the

center of the plat, *A*, or at some other convenient point.
1, 2 and 3 are sheds for the storage of sash, shutters, etc.
If one has but a few frames it will be desirable to have
them seven feet apart, which will give space for the sash
to be drawn off, and will allow a cart to dump its load
of manure or soil between the frames. When a large
number of frames are used, and especially if the land is
valuable, it will be better to have the rows from two to
three feet apart. The shutters and sash can then be
placed in piles at the end of the rows. If only a small
amount of ventilation is needed, the sash can be slipped
up or down, or can be raised, as seen in Fig. 85.

## MAKING THE BED.

For a winter hotbed, the heating material should
have a depth of from two to two and a half feet, while
in the spring one-half that depth will answer. An excavation two feet wider than the frame, and of the required
depth, should be made, in order to prevent the frost
from working into the bed, although for spring use the
same result may be effected by piling fresh manure about
the frame. The heating material is generally fresh
horse manure. Unless it contains a liberal amount of
straw, or similar bedding material, something of the
kind should be added, so that it will not be more than
one-half of clear manure. Oak leaves, also, make a
good material to mix with the manure, as they will hin-
der, and, consequently, prolong the decomposition of
the mass, thus giving an even heat.

About two weeks before the bed is wanted, the
material should be placed in a pile about eight feet wide
and four feet high, with a flat top and vertical sides.
The pile should preferably be made in a shed or manure
cellar, but may be in the open air, or even in the frame
itself. In three or four days it will be fermenting rapidly, and should then be forked over, throwing the out-

side portion to the center; at the end of two or three days the pile should be well warmed up, and the bed may be made, or, if it has not warmed evenly, it should be again turned over, before being placed in the frame. In working over the pile, all coarse lumps should be broken up, and the heap should be left as light as possible, to encourage fermentation. If, when the material is placed in the frame, it is quite warm, it may be leveled off and firmly tramped down, filling it up to within six or eight inches of the glass. Should it not be as warm as is desirable, it may be best to delay the final tramping for a couple of days.

The bed is now ready for the soil, which should be a rich compost. For many crops the soil and manure

FIG. 88.   COLD PIT.

from an old bed will answer. The best materials would be equal parts of pasture sods, decomposed manure, garden soil, and sand enough to make a light mass and prevent baking, spread over the manure to the depth of six inches. For two or three days there will be a violent heat in the bed, but this will soon go down and the bed will be ready for seeds or plants. If, while preparing the manure for the bed, it is found to be dry, it should be moistened with tepid water from a watering can.

In caring for hotbeds, the mats and shutters should be taken off on pleasant days, as soon as the sun is well up, and on bright days the beds should be given air about the same as in a forcing house. The beds should be closed, at least, two hours before sunset, and the covers should be put on as the sun goes down.

While hotbeds are a great convenience after the first of March, they are each year becoming less used for the growing of winter crops. The cost of forcing houses is but little more, and they are much more convenient and in every way more satisfactory.

### DETACHED COLD FRAMES AND PITS.

The most common form of cold frame (a hotbed frame and sash without any heating material) is a low structure used to carry through the winter pansies, violets and other half hardy plants, or for the growing of vegetables and bedding plants, before the danger of frost in the open ground is over in the spring. For many purposes, however, a deep frame or pit is desirable. In Fig. 88 is a cross section of such a structure. If made eight feet wide inside, five or six feet deep at the plate, and of any length, it will be found one of the most useful "rooms" in a greenhouse establishment. The walls may be of wood, brick, stone or concrete, and the top should consist of plates, firmly anchored to the walls, a ridge, rafters and hotbed sash. For a pit eight feet wide inside, the sash on one side should be about six by three feet, and on the other four by three feet, or if ten feet wide, an even span roof can be made, with sash six by three feet on both sides. With mats and shutters, frost can be kept out from a deep pit of this kind. As in Fig. 88, a double use can be made of such a frame. By placing a plank floor about one foot below the plate, the upper portion can be used for violets and similar plants, while bulbs for winter forcing can be plunged in sand upon the bottom. It can also be used for wintering a great variety of plants. In Fig. 10 can be seen a very convenient frame against a greenhouse wall. By means of slides in the wall, a sufficient supply of heat can be admitted from the greenhouse if desired.

## CHAPTER XXVII.

### CONSERVATORIES.

As usually applied, this term refers to small greenhouses attached to dwellings, in which, although plants may be grown, the real object is to have plants shown that are attractive, either in foliage or flower. In a strict sense, however, a conservatory is a structure in which plants that have been developed in narrow and comparatively low greenhouses, known as growing houses, are shown during their period of flower. They may be attached to the dwelling or other building, Fig. 89, but are generally detached buildings surrounded by the various growing houses. In another chapter descriptions will be found of various small structures adapted to the wants of amateurs, but at present we shall consider various combinations of glass structures, consisting of conservatories or show houses, with the necessary subsidiary growing houses, such as would require the care of a professional gardener.

In addition to the large number of public institutions where large conservatories are desirable, the number of private individuals who have the means to erect and maintain establishments of this kind, and taste to appreciate the beauties of the flowers and plant grown

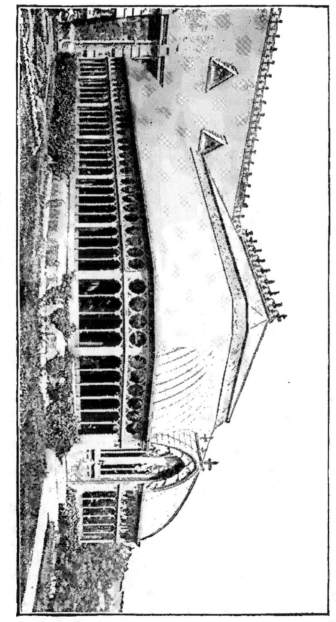

FIG. 89. LARGE ATTACHED CONSERVATORY.

*Erected by Lord & Burnham Co.*

in them, is constantly increasing. The use of plants for purposes of lawn and house decoration, and of flowers for embellishing the table, the parlor, or the person, has become so common that where one can afford it, the possession of a greenhouse has become very desirable, and almost a necessity in some cases.

Conservatories, in proportion to their length, are much wider and higher than the growing houses, and, in fact, there is practically no limit to their size, except the money to erect and maintain them. They are usually erected upon a brick or stone foundation about two and one-half feet high, and with vertical glass sides above this to the height of from six to ten feet above the masonry. The width of the house may vary from twenty or twenty-five feet, to eighty or one hundred and even more, and the length may be as desired. For narrow houses, up to a width of thirty feet, the even span roof with straight rafters, continuous from ridge to plate, will be the least expensive; it will grow the best plants, and, if made in proportion, will not be displeasing.

For a house of this kind the slope of the roof should be about thirty-five degrees. Twenty years ago it was the custom to surmount conservatories of this kind with a lantern top about six feet wide, Fig. 90, two feet high at the plate, and three at the ridge, running the length of the house. These had ventilators in the side walls, which were desirable in summer, but during the winter they added greatly to the consumption of fuel. The lanterns were so narrow that they were of little use, except to add, in a slight degree, to the appearance of the house. They are now no longer used except upon very wide conservatories, where they are so constructed, as shown in Fig. 91, that they add from five to ten feet to the height of the house, and are of such a width that this space can be utilized by tall plants. The straight sash bars can also be used in wide houses, but they will

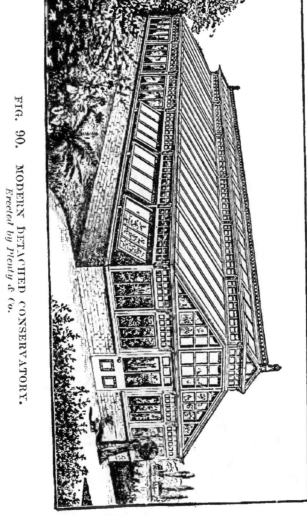

FIG. 90. MODERN DETACHED CONSERVATORY.

*Erected by Plenty & Co.*

have a barn-like appearance, unless the roof is broken by
gables and secondary slopes. This method of building
greenhouses has many things in its favor, that are worthy of commendation. It is in the wide conservatories
that the curvilinear roofs are particularly desirable. In
themselves they are quite ornamental, and they moreover

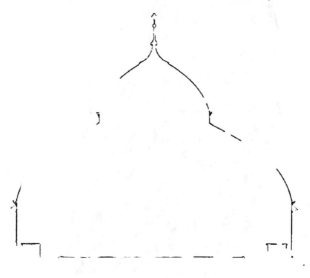

FIG. 91.   CONSERVATORY (section).

form a convenient method of arching over any wide
space, as shown in the cross section, Fig. 91, and in
perspective in Fig. 105.

### IRON HOUSES.

For any structure of this kind, wood is too perish-
able, and the necessary strength could only be secured
by the use of a heavy framework. Of all materials at
present available, architectural iron seems best adapted
for this work, and all sills, posts, rafters, braces, ridges,
purlins and supports should be of this material. There
seems to be but little choice between cypress and metal
sash bars for large conservatories, aside from the larger

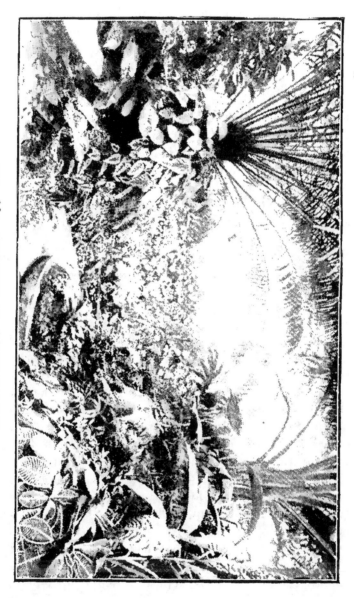

FIG. 92. INTERIOR OF CONSERVATORY.

expense that must be incurred for the latter. If the latter are used, it is desirable that they should have a steel core, as, if constructed of copper, zinc, or galvanized iron, they are likely to bend and crack the glass.

The first cost of the iron roof is considerably more than for cypress, and, in order to be lasting and free from rust, it will need to be painted fully a often. A metallic glazed house is harder to heat than a putty glazed one, and after a year or two is likely to leak heat from, and rain into the house. With the same attention to painting and repairing a wooden roof as is necessary with an iron one, the house will be tighter, easier to heat, there will be less drip, and it will be in a good state of preservation at the end of twenty-five or thirty years. Although practically indestructible, the glazing strips used in the iron houses will have become so bent and out of shape that many of them will require renewal even before this time.

### INTERIOR ARRANGEMENT OF THE CONSERVATORY.

In arranging the interior of the conservatory, it will be well to use all of the center of the building for large palms, bananas, bamboos, tree ferns, and other tall-growing plants, Fig. 92. They should be planted in the ground, and so arranged as to present as natural an appearance as possible. The walks should be of generous widths, and so arranged as to bring into view all parts of the house. The portion of the house next to the walls may be arranged in the same manner as the center, but it is desirable to have a portion of it, at least, supplied with tables, upon which plants in flower may be displayed. If they are combined with ferns and ornamental-leaved plants the effect will be very pleasing. This is really the purpose of a conservatory, since, as is usually the case, if it is kept at a temperature of fifty-five to sixty degrees when plants are brought in from

FIG. 93.    INTERIOR OF A PALM HOUSE.
*Pitcher & Manda, Short Hills, N. J.*

the stove and other warm rooms, the flowers will be con-
served, and will last must longer than if kept at a high
temperature. Frequently the large rooms are used for
growing collections of the more ornamental palms, and
are known as palm houses, Fig. 93.

### THE STOVE HOUSE.

As first used, the term "stove" was applied to
greenhouses in which artificial heat was supplied by
means of stoves. As is frequently the case, the name of
the object became attached to the building in which it
was used, and a stove house to-day is merely a hothouse
with a temperature of sixty-five to seventy-five degrees.
As a rule, these are considerably narrower and lower
than the conservatories or palm houses. They are sel-

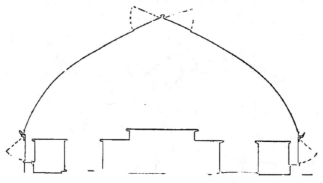

FIG. 94.    STOVE ROOM (*Section*).

dom wider than twenty or twenty-five feet, and from
twelve to twenty feet in height. If built upon a masonry foundation two and a half feet high, the vertical
side walls are usually about two and one-half or three
feet high, with side ventilators. The roof has an angle
of thirty to thirty-five degrees, with ventila or on each
side of the ridge.

This house of the more stove house. The center table may be used, or if the plants are large, they

FIG. 95. COMBINED STOVE AND ORCHID HOUSE.

may be planted or plunged. In Fig. 94 is seen a cross section of a stove house with a curvilinear roof, while in Fig. 95 an interior view of the same house is seen.

When one does not desire the curvilinear roof for itself, a stove room built with straight sash bars will give fully as good results. A very pleasing effect may be produced when stove plants and orchids are grown in the same room. So far as the construction of the house itself is concerned, a stove house does not differ from others of the same general style, except that to obtain the proper temperature, the radiating surface, provided in the steam or water pipes, must be considerably larger than for most houses.

## COOL HOUSES.

In all establishments of this kind there should be, at least, one house in which a maximum night temperature of fifty degrees is maintained, for such plants as do not require the stove room heat. In a general way, their construction would be the same as for a stove house, although, as a rule, a narrower house will answer. Where many bedding plants are used for lawn decoration in the summer, a similar house will be required for that purpose. If desired, a portion of this room could be used for propagating purposes, or a narrow house could be erected especially for propagation.

When large palms, and other similar plants, are used upon the lawn during the summer, they should be stored in a cool house, and if no other place is at hand, a lean-to against a shed or other building can be cheaply erected, and a proper temperature can be maintained at very little expense. For many of the broad-leaved evergreens, that should be kept in a dormant condition during the winter, a north side lean-to house is quite desirable

### ORCHID HOUSES.

As indicated above, orchids can be grown in stove houses with other plants, and for many amateurs no special orchid house need to p ted, but when the collections are large, it will be w l to have houses set apart for their use. It is genera ac itted that no form of construction is better adapte r orchid culture than the span roof house. Many g w rs have made the mistake of erecting high and wide n ms s, while, had they confined themselves to structur not over sixteen or eighteen feet wide, and ten or clev n feet high, they

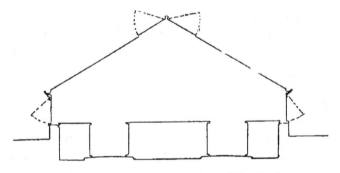

FIG. 96. ORCHID HOUSE (*Section*).

would have obtained more satisfactory results, to say nothing of the loss in cost of construction and fuel.

The orchids are divided int three groups,—stove, intermediate and cool house,—from the temperature in which they thrive best, and houses should be provided accordingly. If an orchid house sixty to seventy-five feet long is erected, it can be divided by cross partitions into three rooms, which can be adapted for the different classes of orchids by a proper adjustment of the heating pipes.

For small plants, a house only twelve feet wide and eight feet high at the ridge will be even easier to erect and hea , but the moisture and temperature cannot be

12

controlled as well as in a wider house. For some of the cool house orchids, a lean-to house answers quite well, and where Cattleyas are grown in large quantities for market, the three-quarter span house will give good satisfaction.

In Fig. 96 will be found a section of an orchid house, showing the arrangement of the tables and ventilators. At least two feet of the side walls should be above the masonry, giving sixteen or eighteen inches of glass. There should be two lines of ventilators at the ridge, and some means of bottom ventilation should also be provided. In the intermediate and Mexican houses the vertical sash in the side walls may be used as ventilators, but in the stove or East Indian house all drafts of cold air are injurious, and it is preferable to admit the fresh air under the tables.

All orchids require very careful shading during the summer. A thin permanent shading may be given in the spring, but the main reliance should be upon blinds or curtains of canvas or netting, that can be drawn up except while the sun is shining bright. In dull weather a thick permanent shading would be injurious to the plants.

### GRAPERIES.

The large, choice varieties of European grapes are not hardy in our latitude, and some protection must be provided for them if we are to grow them. Many varieties can be grown in a glass house even without heat, and to such a building the name of "cold grapery" has been given. Some varieties require heat to bring them to maturity, while others can be brought in quite early if started in winter with artificial heat, and for such purposes the "hot grapery" is used, although the name "forcing grapery" is also applied to it.

While almost any greenhouse will answer for growing grapes, experience has shown that certain forms are

better than others. For the forcing grapery nothing
seems to be better than the narrow lean-to or two-third
span, such as is seen in cross section in Fig. 97, as it
furnishes a warm back wall against which the vines can
be trained, and, like all lean-to houses, it is cheaply constructed and heated. The three-quarter span houses are
also excellent for either forcing or cold graperies, but
unless one has walls that can be used for this purpose it
will be preferable to build span roof houses running

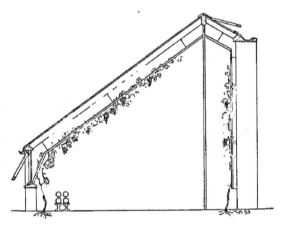

FIG. 97. FORCING GRAPERY (*Section*).

north and south, except when grapes are to be forced in
the winter. The span roof house, Fig. 98, encloses a
larger body of air than either of the other houses, and it
will be easier to regulate the temperature and the moisture in such a house than in a narrow one. In addition
to the above reason the wide houses are preferable, as
they have longer rafters and afford more space for training the vines.

The curvilinear roof is frequently used for vineries,
and in Fig. 99 is shown a section of a curvilinear house.
They give somewhat longer rafters for training the vines,
but they have no other advantage, except, perhaps, in

appearance, and this will not counterbalance the increased cost.

The even span greenhouses, with straight sash bars, seem to be the favorite form for graperies. In their general construction they do not differ from even span structures of similar dimensions used for other purposes, and for the details of construction reference is made to Chapters V and VI. There are, however, certain points that should be considered in erecting a grapery. If the

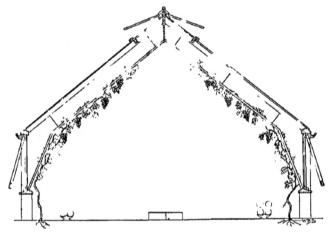

FIG. 98.   EVEN SPAN GRAPERY (*Section*).

house is a wide one, the slope of the roof may be less than if it is comparatively narrow, and thirty-five, or even thirty degrees pitch will be sufficient in one case, while forty, or perhaps forty-five degrees, may be desirable in another.

In choosing a site for a grapery, it is well to have it somewhat sheltered from the north and east, and, by all means it should be well drained to the depth of three feet, that the border may not become wet. The situation should be such that it will not be affected either by the shade, or the roots of large trees, which might get into the border and steal from the vines.

The wall of brick or stone, if either be used, should extend for a foot or so above the level, and if a portion of the border is to be on each side of the wall, arches should be left in the wall at intervals of about three feet, with openings at least one foot square through which the roots can make their way. Upon this wall there should be one of wood two feet high, with continuous side ventilation (see Fig. 98). If it is desired to make the first cost as low as possible, the side walls may be built of wood, without the use of a stone foundation, but in the damp border it will not be very durable, and the form of wall described above is preferable. The roof

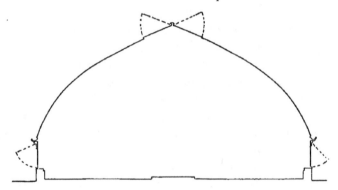

FIG. 99. CURVILINEAR GRAPERY (*Section*).

should be somewhat stiffer than for an ordinary greenhouse, but it need not be different in construction from those described in Chapter VI. There should be, at least, one line of ventilation at the ridge, and preferably two in a wide house.

While steam could be used for heating vineries, it has not to any extent, hot water being relied on for the most part. The flue is not satisfactory and is but little used. In arranging the pipes, it is best to have them three or four feet from the vines, as, if in close proximity, they might unduly dry out the border, and would tend to invite the development of the red spider. The

radiation should be ample, and so supplied with valves
that the amount of heat furnished can be regulated at
pleasure.

Some arrangement should be made for training the
vines, and perhaps the simplest form of trellis will be
made of No. 12 galvanized wires, arranged one foot
apart, and suspended about fifteen inches below the
sash bars (Figs. 97 and 98).

### ORCHARD HOUSES.

In many sections of the country some of our
choicest fruits, such as peaches, nectarines, apricots,
sweet cherries, etc., cannot be grown in the open air,
and if their cultivation is attempted it must be under
glass. In Europe fruit houses are very common, and
for many years have formed an important part of the
greenhouses, not only upon the large estates, but in connection with the cottages of the middle classes. On this
side of the Atlantic, the ease with which these crops can
be grown in favorable localities, and the abundance and
cheapness of the sub-tropical fruits from Florida and
California, have united to restrict the use of orchard
houses. While it is doubtful if they can be made profitable as commercial ventures, except under unusually
favorable conditions, many persons find them very desirable to furnish a supply of fresh fruit out of season for
their own tables.

In their construction, orchard houses do not greatly
differ from graperies. The walls are built in the same
manner, but should have a height of six feet, at least
one-half of which should be of glass. They may be
constructed of wood or iron posts and boards up to a
height of two feet, or they may have a masonry foundation with a brick wall above. The glass in the side walls
should be from three to four feet high, and at least one-
half of it should be in the form of ventilators, hinged at

the top. The roof may be either of movable sash or of fixed sash bars, the latter being preferable if the house is to be a permanent one. In this case iron posts, rafters, purlins and ridge, with cypress sash bars, can be used to advantage. With the high walls, to give room for the trees at the sides of the houses, it will be desirable, particularly if the house is a wide one, to give the roof a comparatively low pitch, in order to bring the glass down as near as possible to the plants in the center of the house. A slope of twenty-six degrees will answer, and if the house is more than twenty-five feet wide, twenty degrees will be preferable. Ample means of ventilating the houses should be provided. In addition to the row under the plate, there should be one, in narrow houses, and two in wide ones, at the ridge, and doors or ventilators in the ends are also desirable.

While any form of house, lean-to, even span, or three-quarter span, curvilinear or straight, may be used, the wide even span will be most satisfactory, as the light will be more evenly distributed than in either of the other forms, and the temperature and moisture will be easier to regulate than in a lean-to or in a narrow house. While houses not over twelve or fifteen feet wide will give fair results, a width of eighteen, twenty, or, better yet, twenty-five feet will be preferable. The lean-to will be a cheap form to erect, when it can be built against the south wall of a building, but when this cannot be done, an even span house will cost no more, and will be much more satisfactory. With a lean-to construction, a house about fifteen feet wide can be built when the south wall is five or six feet high and the north one fourteen or fifteen feet. While not really desirable, a narrow lean-to house six to ten feet wide can be used. The construction would be about the same as that of a narrow lean-to grapery. In this kind of a house the trees are generally trained upon the north wall. With proper care in regulating the heat and moisture, and in ventilating, fair results will be obtained. If to be used as fruit-forcing houses, the three-quarter span house, with the long slope either to the north or to the south, can also be used.

While some growers plant the trees in the border, others grow them in pots or boxes, and are then able to pack the trees away, and use the house for other purposes until it is necessary to start the trees in the winter or spring. In narrow houses there is only one walk, the trees being arranged upon either side, but in wide span roof houses, although this arrangement is often made, it is preferable to have two walks, one on either side, about four feet from the walls, thus securing the peak of the roof as an additional space for tall trees. When there is a walk in the center of houses over fifteen feet wide, it is necessary to have a narrow walk upon either side, for convenience in watering and caring for the plants.

## FIRE HEAT.

Even when only used as growing houses, it is desirable to have the houses provided with heating apparatus. While dormant it frequently happens that the temperature may drop so low that the buds will be injured, since, as a rule, the buds are not as well ripened as when grown in the open air, and will be more susceptible to cold. It is after the buds start, however, that the danger of injury by cold is greatest, as, if the temperature falls below the freezing point while the trees are in bloom the crop will be lost and the trees greatly injured.

Although steam or hot air flues may be used for heating, hot water will be found more satisfactory. The piping should be sufficient to keep the temperature of the house at forty five degrees in the coldest weather that is likely to occur after the trees are started. If peaches or other fruits are to be forced, they should be started as soon as March 1, and the forcing may commence as early as January. In the forcing house a temperature as high as fifty or fifty-five degrees at night is necessary for the best results, and in estimating the radiation this should be kept in mind.

## CHAPTER XXVIII.

### THE ARRANGEMENT OF GREENHOUSES.

When a large number of houses that are used for different purposes are to be combined, considerable skill is necessary in order to secure the best results. The arrangement depends largely upon the kind of houses, as well as their size and shape, and as there are, at least, ten or a dozen houses that go to make up a complete plant, in the very selection of the houses for the establishment there would be opportunity for hundreds, and even thousands, of combinations. We have outlined above the structural peculiarities of seven or eight of the more important houses, and have elsewhere described the rose house, propagating house, forcing house, etc., and now offer for consideration perspective views and ground plans of greenhouse establishments, designed by the leading horticultural architects and builders of the country.

The first illustration, Fig. 100, shows a part of the greenhouses at the Michigan Agricultural College, and in Fig. 101 a ground plan of the houses is shown. This range of houses is not an elaborate one, but it is well arranged, and in connection with a grapery and two forcing houses makes a fairly complete establishment. It contains a palm house, or conservatory, fifty-eight by twenty-five feet, a stove room twenty-five feet square,

FIG. 100. SMALL RANGE OF GREENHOUSES.
A portion of the greenhouses at the Michigan Agricultural College.

a cool house of the same size, a rose room eighteen by twenty-five feet, two propagating (hot and cold) houses for the growing of bedding plants, each twelve by fifty feet, and another room twenty-five by twenty-four feet, that is used as occasion demands. The workroom is twenty-five by fifteen feet, and is over the heaters. The gardener's house, as shown in the illustration, is joined to the conservatory.

The houses shown in perspective were erected in 1892, by Lord & Burnham Co., and are of their iron

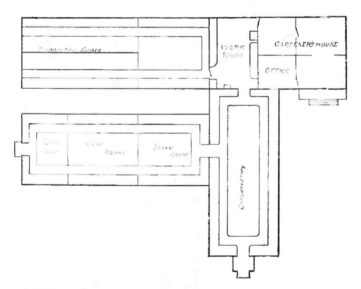

FIG. 101. GROUND PLAN OF MICHIGAN AGRICULTURAL COLLEGE GREENHOUSES.

frame construction, with all outside work of cypress. The walks are of cement, the tables of angle iron with gas pipe legs, and with slate tops in some rooms and tile in others. The houses are heated by a No. 8 Furman hot water heater, put in by the Herendeen Manufacturing Co., of Geneva, N. Y. The method of constructing the walls, roof, benches, and the heating coils is shown

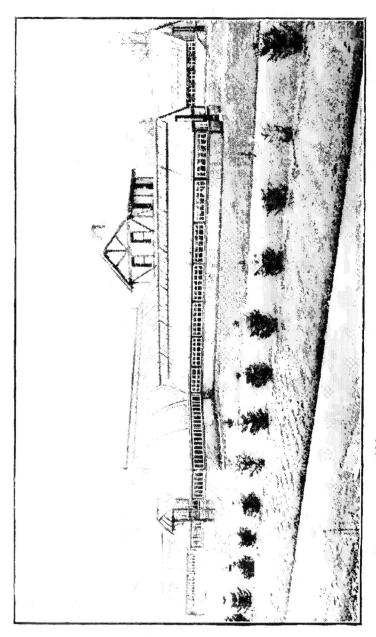

FIG 102.   CONSERVATORY AND GROWING HOUSES.

*Erected by Thos. W. Weatherell's Sons.*

in Figs. 14, 17 and 26, except that the coils here used are double. Each room is piped independently, and each coil is so arranged that heat can be shut off in whole or in part from one room without affecting the others. In the rear of the new greenhouses, as shown in the ground plan, are three other houses, that were erected some fifteen years ago. They are entirely constructed of wood, and although kept well painted, are showing signs of decay in some places. They are heated by a Spence hot water heater, and the radiating surface is supplied by two-inch flow pipes and one and one-half inch returns.

Similar in construction, in many respects, to the new greenhouses described above, are those shown in Fig. 102. The principal difference is in the form of the roof of the conservatory, which is curvilinear, and in the arrangement of the growing houses. This range was erected by Thos. W. Weathered's Sons, at New Dorp, Staten Island.

A larger and more expensive range of houses is shown in Fig. 103. The ground plan of these houses is illustrated in Fig. 104, and from this the size and uses of the different rooms can be ascertained. The expense of such a range of houses, complete with iron tables and heating apparatus, will not be far from $16,000. They were erected by Hitchings & Co., of New York City. The range, as will be seen, consists of an elongated hexagonal palm house with a curvilinear roof. The side walls are quite high, and, with the pitch of the roof, affords room for growing quite large plants. From each side of the conservatory, facing east and west, extend two span roof growing houses, which can be used for stove house, cool stove, and hot and cold propagating houses, or houses for carnations and other flowering plants. At the end of the north houses will be seen a long three-quarter span rose house and a large work-

FIG. 103. PERSPECTIVE VIEW OF HITCHINGS & CO.'S RANGE OF GREENHOUSES.

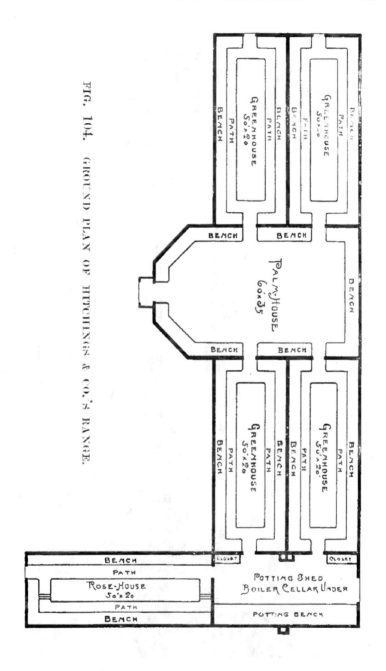

FIG. 104. GROUND PLAN OF HITCHINGS & CO.'S RANGE.

FIG. 105. EXTENSIVE RANGE OF HOUSES.
*Erected by Lord & Burnham Co.*

room.  The method of construction used by Hitchings
& Co. is not particularly different from the one used by
Lord & Burnham Co., the principal difference being
that the former generally make the rafters and posts in
separate pieces, which are clamped together by an iron
bracket at the plate, and use iron gutters and eave

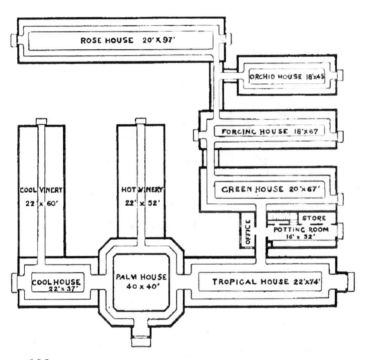

FIG. 106.   GROUND PLAN OF LORD & BURNHAM CO.
RANGE.

troughs, while the latter use wooden gutters, and forge
the posts and rafters from one piece.

If anything more elaborate is desired, it can be
found in the range shown in Fig. 105, the ground
plan of which can be seen in Fig. 106.  It was designed
and erected by Lord & Burnham Co., at Yonkers, N. Y.,
and, as will be seen from the illustrations, consists of an

13

octagonal curvilinear conservatory forty by forty feet, which is shown in cross section in Fig. 91; a stove or tropical house, with a curvilinear roof, twenty-two by seventy-four feet, Fig. 94; a cool house twenty-two by

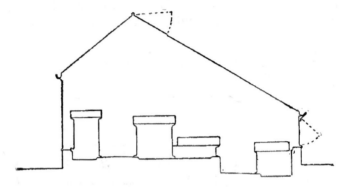

FIG. 107.   FORCING HOUSE (*Section*).

thirty-seven feet, also curvilinear; a span roof green-house twenty by sixty-seven feet; a forcing house, with a three-quarter span roof, eighteen by sixty-seven feet,

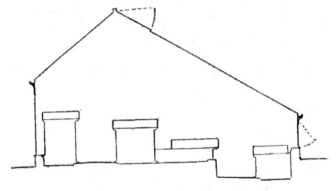

FIG. 108.   ROSE HOUSE (*Section*).

Fig. 107; an even span orchid house, Fig. 96, eighteen by forty-five feet; a three-quarter span rose house, Fig. 108, twenty by ninety-seven feet; a cool vinery, Fig. 99, twenty-two by sixty feet, and a hot vinery, both curvilinear, twenty-two by fifty-two feet, besides a potting room and office.

This establishment is very complete, and seems to be well arranged. If plain curvilinear houses are desired, the forms shown here have been thoroughly tested, and have been found quite satisfactory.

## CHAPTER XXIX.

### GLASS STRUCTURES FOR AMATEURS.

Many lovers of gardening, who are restrained from indulging in their favorite pastime by our long six months of winter, would gladly erect small glass structures in which to prosecute many of the lighter operations of gardening, but are deterred by what they imagine to be the excessive cost. In this chapter an attempt will be made to outline methods of constructing several forms of small greenhouses that can be cheaply erected, and which will be found very useful and entirely satisfactory. Upon many town, as well as country places, we often find small cold frames, or cold pits, in which half hardy plants can be stored through the winter, and in which many of the hardier vegetable and budding plants can be started and grown. At best, they are of little value in forwarding plants during the severe parts of the winter, and are far from satisfactory in every way.

For our present purpose a house is needed, perhaps ten by fifteen feet in area, convenient to or attached to the dwelling, that can be attended to in all weathers without exposure, and that can be cheaply constructed and maintained.

### ATTACHED CONSERVATORIES.

It is frequently desirable to have, in connection with the dwelling, a room enclosed with glass, in which

flowers can be grown or exhibited.   The large structures
that are sometimes seen do not differ, in their principal
features, from detached conservatories, and need no consideration here.   Although the best results, so far as
the growth of the plants is concerned, cannot be obtained
in a lean-to structure, the fact that small conservatories
can be placed in an angle of the dwelling, where the
walls of the house will form the end and rear of the con-

FIG. 109.   VERANDA CONSERVATORY.

servatory, and thus greatly reduce the cost of construction, leads to their use when the first outlay is considered.   These attached conservatories, in the lean-to
style, may vary in width from six to fifteen feet, but if
anything wider than this is desired, it will be best to
have detached houses, or to use some other form of roof.
     The simplest kind of a conservatory of this style is
made from an ordinary veranda, in which the spaces

between the side posts are filled in with glass sash.
These can be taken out in the summer, if desired, and
the veranda restored to its ordinary use. By the addition of a glass roof, far better results can be obtained,
however, and if a veranda conservatory is to be built, it
will be found cheaper than a wooden or a tin roof. It
should be eight or nine
feet wide, to secure the
best results, although a
veranda five feet wide
will answer as a con-
servatory.

In constructing
these veranda conserva-
tories, Fig. 109, a loca-
tion on the south side
of the house should be
selected, as a rule, al-
though for ferns and
similar plants, the east,
or even the north side
is preferable. The
framework of the con-
servatory should be put
up in a permanent man-
ner, as should the en-
tire roof, but it will be
sometimes found best,
if the glass in the side

FIG. 110. VERANDA CON-
SERVATORY (*Section*).

and ends is of temporary sashes, so arranged that they
can be taken out, as seen in cross section, Fig. 110. The
floor should be at the same height, and constructed in
the same manner as for an ordinary veranda, although a
cement floor may be used if desired. In case a wooden
floor is used, the veranda should be closed in below, or
ceiled against the floor joists.

If designed as a conservatory for flowers, a doorway in the wall of the dwelling should be arranged in the middle, either of the side or end, and in case the conservatory is a large one, it will be convenient to have an outer door. As a rule, these doors should be opposite each other. It is also an excellent plan to have the portion of the wall of the house adjoining the conservatory, of glass. The posts should be from five to seven feet high, and placed five feet six inches apart. At the height of two feet a sash sill should be placed, and the space beneath should be filled in to correspond with the finish of the house. The walls of the veranda above this sill may be of permanent sash bars and glass, or, as is better, unless it is to be used as a conservatory throughout the year, the spaces between the posts may be filled in with glass sash that can be taken out during the summer.

If a veranda is made eight feet high at the eaves, this will admit of the placing of a ventilating sash in the front wall, but in low structures it will have to be placed in the roof. The conservatory roof should have rafters of two by four inch cypress running from each post to the wall of the house. The remaining framework of the roof will consist of two by one and one-eighth inch sash bars. In Fig. 110 is shown a cross section of a house six feet wide, from which the details for the construction of the walls and roof can be ascertained, while Fig. 109 gives an idea of the exterior appearance of the same conservatory.

If the amount of glass exposed is not too large, necessary heat can be supplied from the adjoining living room for so-called cool house plants, but it will be desirable to have heat directly supplied to the room. Hot water or steam heating pipes, arranged as in Fig. 110, will be desirable, but if nothing better is available, one or two large kerosene heating stoves can be used, pro-

FIG. 111.  SMALL ATTACHED CONSERVATORY.
*Erected by Lord & Burnham Co.*

vided pipes are arranged to carry off the gases of combustion.

If more elaborate structures are desired, they should be of a style that will correspond with that of the residence. The location at the corner of the house, as seen in Fig. 111, is desirable, as light can be obtained from three sides, and better results can be obtained than in a veranda conservatory.

## DETACHED GREENHOUSES FOR AMATEURS.

It frequently happens that for some reason it is not desirable to have the conservatory attached to a building, in which case there will be a great variety of structures from which to select. Here, again, the lean-to form will be found a cheap one to erect, and the same directions apply here as in a large house. In Fig. 112 is presented a cross section of a house built by Chas. Barnard, and described in the *American Garden*, October, 1890. The walls were built as shown in the engraving, sheathed on both sides, and with a layer of hair felt inside. A cheaper wall could be erected by double boarding the outside, and having a layer of heavy

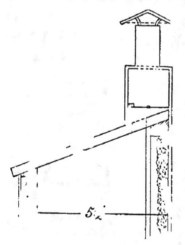

FIG. 112. A CHEAP HOUSE. (*Section*).

sheathing paper between the boards. The outside boarding of the north wall extends one foot above the roof, to act as a wind-break. The walls measure five feet six in be outside, and the roof is formed of hotbed sash, the joint being made tight with battens. Ventilation is secured through a cupola in the center of the ridge.

The base of this is twelve inches square, and the circulation of air is controlled by a damper, as will be seen from the engraving. If other forms of houses are desired, they will only be miniatures of the large ones described in previous chapters.

## PORTABLE CONSERVATORIES.

Several builders make a specialty of supplying houses of the kind. Fig. 113 shows one of these houses put up by Hitchings & Co., and in Fig. 114 is seen the same house with a portion of the sash removed. As will be seen, the houses are built with an iron frame, similar to that used in large houses, and covered with sash that can be very quickly put in place. They are supplied with hot water heating apparatus, and ventilating machinery. Besides being portable, the houses are extensible, and another section can be added with little trouble at any time. A house eight by sixteen feet, with heating and ventilating apparatus, costs about $360. It makes a very durable house, and is, in every way, first class.

If one cannot afford so expensive a house, a very satisfactory conservatory can be built by using four by four inch posts for the walls, set four feet apart, and with every other post four feet high, the others being cut off at the height of two and one-half feet; sash, sills, plates, end rafters and ridge pieces can be obtained, cut in the desired shapes, at any wood-working factory, or from dealers in greenhouse materials. The roof may be made of fixed sash bars or of temporary sash.

The heating apparatus for a narrow house will cost from four to five dollars per linear foot, and the ventilating apparatus from ten to fifty cents, according to the kind used. The lumber can be estimated at about $3.00 per linear foot, and the glass will not be far from $1.50 per foot, while the labor of carpenters and painters will

FIG. 113. PORTABLE IRON CONSERVATORY (*Perspective*).

FIG. 114. PORTABLE IRON CONSERVATORY (*Section*).

be about $2.50 per foot, or a total of $225 to $250 for a substantial house fifteen by twenty feet, with heating apparatus and benches.

### THE BASEMENT PIT.

There are some objections to the structure to which the above name has been given, the principal ones being that it is somewhat difficult of access, that it is inconspicuous, and that the plants grown there do not give the pleasure they would, were it entirely above ground

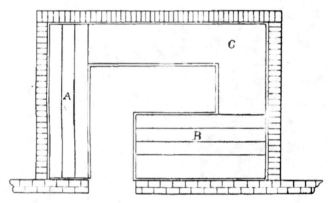

FIG. 115.   GROUND PLAN OF BASEMENT PIT.

and separated from one of the living rooms by a glass partition.   The points claimed for the basement pit are, that it is very easily heated, owing to the comparatively small area of exposed glass, and its cheapness of construction, and every one of the objections urged against it could be overcome by raising the structure to the ground level and supplying the needed glass partition and door.

For the location, it is best to select the south side of the dwelling, if possible, although the west, or east, or a point between would answer.   If it can be situated where a cellar window is located, all the better.   The wall is torn away so as to afford an opening three feet

wide, in which a glass door should be placed. The excavation is then made, and the wall of masonry laid up
to the general level outside (Fig. 115). The highest
point of the roof should be located four feet above the
top of the foundation. Four by six inch sills, Fig. 116,
(*a*) should then be put down, and upon these four by
four inch posts (*b*) twenty inches high, and with rabbets
for glass, should be placed at the corners and at intervals of five feet along the side
and ends. The two by five
inch plate is then placed, as
seen at (*c*), and a four by two
inch fascia (*d*) should be set
into the tops of the posts
along the front, and two by
four inch rafters (*f*) should
extend from the top of the
posts to the wall of the house,
where they should be sup-
ported by a four by one inch
ridge board. A gutter (*e*), if
desired, will complete the exterior of the structure, with
the exception of the sash bars,
purlins, ventilators, and glaz-

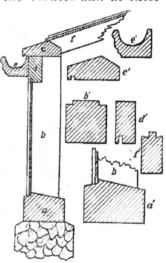

FIG. 116. DETAILS FOR
BASEMENT PIT.

ing, which will not differ from the same parts of a greenhouse. The interior arrangement will depend upon the
uses to which it is to be put. The doorway may be in
the center of the back wall, or, better yet, about three
and one-half feet from the east end. If desired, a flat
table (*C*) three feet wide, for starting cuttings and seed-
lings, and for the growing of vegetable and bedding
plants, can extend along the west and south sides, and
on the east (*A*) and north (*B*) sides tiers of shelves may
be placed, on which the larger plants can be arranged.
The amount of heat required for such a house will

depend considerably upon the kinds of plants to be grown, as for many greenhouse plants a temperature of forty-five to fifty degrees is ample at night, and if it occasionally drops below forty degrees no harm will be done, while most of the so-called stove plants would be injured if the temperature remains below sixty degrees for any length of time. If the pit opens into a large cellar, particularly if it contains a hot air furnace, or other heating apparatus, there will be little danger of frost, except in severe cold weather, when an oil stove placed in the room will be all that is necessary. For

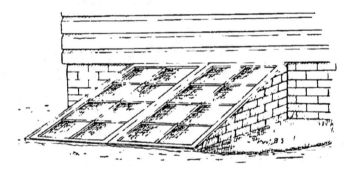

FIG. 117. CELLAR-WAY CONSERVATORY (*Perspective*).

stove plants, the use of a second stove in severe weather would probably be required. If the dwelling is heated by hot water or steam, of course the matter of heating would be very simple, and, if not, the next most satisfactory plan will be to use a small hot water heater, or to obtain heat by placing a coil in a hot air furnace and connecting it with hot water heating pipes in the conservatory. Directions for arranging the pipes, etc., will be found in the chapter on Heating Greenhouses.

Where only a small cold frame or hot bed is wanted, it can be arranged just outside a cellar window. If a frame four by two feet is sunk in the ground and covered with a hotbed sash, the pit thus made can be used for growing quite a variety of greenhouse plants, and for starting plants in the spring. In severe weather a light covering may be required to keep out frost, although if the cellar contains a furnace this will not be necessary, in ordinary winters, except in the colder sections of the country. A writer in *American Garden* describes a cold

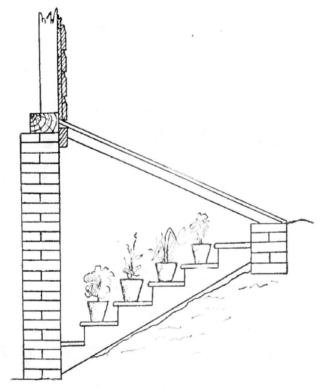

FIG. 118. CELLAR-WAY CONSERVATORY (*Section*).

pit arranged in the outside cellar-way of a house. The doors should be removed and replaced by hotbed sash, as in Fig. 117, which shows the appearance of the cellar-way from the outside. The stairs, Fig. 118, can be used as shelves for the plants, and it will make a good place for wintering any half-hardy plants. By the use

of shutters and mats the frost can be kept out, even
without opening the inner doors.   If artificial heat can
be provided, a variety of plants can be grown.

# INDEX.

## Mushrooms. How to Grow Them.

For home use fresh Mushrooms are a delicious, highly nutritious and wholesome delicacy; and for market they are less bulky than eggs, and, when properly handled, no crop is more remunerative. Anyone who has an ordinary house cellar, woodshed, or barn can grow Mushrooms. This is the most practical work on the subject ever written, and the only book on growing Mushrooms ever published in America. The whole subject is treated in detail, minutely and plainly, as only a practical man, actively engaged in Mushroom growing, can handle it. The author describes how he himself grows Mushrooms, and how they are grown for profit by the leading market gardeners, and for home use by the most successful private growers. The book is amply and pointedly illustrated, with engravings drawn from nature expressly for this work. By Wm. Falconer. Is nicely printed and bound in cloth. Price, post-paid .................................................. 1.50

## Allen's New American Farm Book.

The very best work on the subject; comprising all that can be condensed into an available volume. Originally by Richard L. Allen. Revised and greatly enlarged by Lewis F. Allen. Cloth, 12mo... 2.50

## Henderson's Gardening for Profit.

By Peter Henderson. New edition. Entirely rewritten and greatly enlarged. The standard work on Market and Family Gardening. The successful experience of the author for more than thirty years, and his willingness to tell, as he does in this work, the secret of his success for the benefit of others, enables him to give most valuable information. The book is profusely illustrated. Cloth, 12mo... 2.00

## Fuller's Practical Forestry.

A Treatise on the Propagation, Planting, and Cultivation, with a description and the botanical and proper names of all the indigenous trees of the United States, both Evergreen and Deciduous, with Notes on a large number of the most valuable Exotic Species. By Andrew S. Fuller, author of " Grape Culturist," " Small Fruit Culturist," etc. 1.50

## The Dairyman's Manual.

By Henry Stewart, author of " The Shepherd's Manual," "Irrigation " etc. A useful and practical work by a writer who is well known as thoroughly familiar with the subject of which he writes. Cloth, 12mo ............................................................... 2.00

## Truck Farming at the South.

A work giving the experience of a successful grower of vegetables or garden truck for Northern markets. Essential to any one who contemplates entering this promising field. A. O. Oemler of Georgia. Illustrated. Cloth, 12mo ........................... 1.50

## Harris on the Pig.

New edition. Revised and enlarged by the author. The points of the various English and American breeds are thoroughly discussed, and the great advantage of using thoroughbred males clearly shown. The work is especially valuable to the farmer who keeps but few pigs, and to the breeder of an established herd. By Joseph Harris. Illustrated. Cloth, 12mo .............................................. 1.50

## Jones's Peanut Plant. Its Cultivation and Uses.

A practical book, instructing the beginner how to raise good crops of Peanuts. By B. W. Jones, Surry Co., Va. Paper Cover,..... .50

### Barry's Fruit Garden.

By P. Barry. A standard work on fruit and fruit-trees; the author having had over thirty years' practical experience at the head of one of the largest nurseries in this country. New edition, revised up to date. Invaluable to all fruit-growers. Illustrated. Cloth, 12mo. 2.00

### The Propagation of Plants.

By Andrew S. Fuller. Illustrated with numerous engravings. An eminently practical and useful work. Describing the process of hybridizing and crossing species and varieties, and also the many different modes by which cultivated plants may be propagated and multiplied. Cloth, 12mo ................................................. 1.50

### Stewart's Shepherd's Manual.

A Valuable Practical Treatise on the Sheep, for American farmers and sheep growers. It is so plain that a farmer, or a farmer's son, who has never kept a sheep. may learn from its pages how to manage a flock successfully, and yet so complete that even the experienced shepherd may gather many suggestions from it. The results of personal experience of some years with the characters of the various modern breeds of sheep, and the sheep-raising capabilities of many portions of our extensive territory and that of Canada—and the careful study of the diseases to which our sheep are chiefly subject, with those by which they may eventually be afflicted through unforeseen accidents—as well as the methods of management called for under our circumstances, are here gathered. By Henry Stewart. Illustrated. Cloth, 12mo.... 1.50

### Allen's American Cattle.

Their History, Breeding. and Management. By Lewis F. Allen. This Book will be considered indispensable by every breeder of live stock. The large experience of the author in improving the character of American herds adds to the weight of his observations, and has enabled him to produce a work which will at once make good his claims as a standard authority on the subject. New and revised edition. Illustrated. Cloth, 12mo ........................... 2 50

### Fuller's Grape Culturist.

By. A. S. Fuller. This is one of the very best of works on the culture of the hardy grapes, with full directions for all departments of propagation, culture, etc. with 150 excellent engravings, illustrating planting, training, grafting, etc. Cloth, 12mo ..................... 1.50

### White's Cranberry Culture.

CONTENTS:—Natural History.—History of Cultivation.—Choice of Location.—Preparing the Ground.—Planting the Vines.—Management of Meadows.—Flooding—Enemies and Difficulties Overcome.—Picking.—Keeping.—Profit and Loss.—Letters from Practical Growers.—Insects Injurious to the Cranberry. By Joseph J. White. A practical grower. Illustrated. Cloth. 12mo. New and revised edition 1.25

### Herbert's Hints to Horse-Keepers.

This is one of the best and most popular works on the Horse in this country. A Complete Manual for Horsemen, embracing: How to breed a Horse; How to Buy a Horse; How to Break a Horse; How to Use a Horse; How to Feed a Horse; How to Physic a Horse (Allopathy or Homœopathy); How to Groom a Horse; How to Drive a Horse; How to Ride a Horse, etc. By the late Henry William Herbert (Frank Forester). Beautifully Illustrated. Cloth, 12mo... 1.75

### Henderson's Practical Floriculture.

By Peter Henderson. A guide to the successful propagation and cultivation of florists' plants. The work is not one for florists and gardeners only, but the amateur's wants are constantly kept in mind, and we have a very complete treatise on the cultivation of flowers under glass, or in the open air, suited to those who grow flowers for pleasure as well as those who make them a matter of trade. The work is characterized by the same radical common sense that marked the author's ' Gardening for Profit," and it holds a high place in the estimation of lovers of agriculture. Beautifully illustrated. New and enlarged edition. Cloth, 12mo ............................... 1.50

### Harris's Talks on Manures.

By Joseph Harris, M. S., author of " Walks and Talks on the Farm," "Harris on the Pig." etc. Revised and enlarged by the author. A series of familiar and practical talks between the author and the deacon, the doctor, and other neighbors, on the whole subject of manures and fertilizers ; including a chapter specially written for it by Sir John Bennet Lawes, of Rothamsted, England. Cloth, 12mo ............ 1.75

### Waring's Draining for Profit and Draining for Health.

This book is a very complete and practical treatise, the directions in which are plain, and easily followed. The subject of thorough farm drainage is discussed in all its bearings, and also that more extensive land drainage by which the sanitary condition of any district may be greatly improved, even to the banishment of fever and ague, typhoid and malarious fever. By Geo. E. Waring, Jr Illustrated, Cloth 12mo.
1.50

### The Practical Rabbit-Keeper.

By Cuniculus. Illustrated. A comprehensive work on keeping and raising Rabbits for pleasure as well as for profit. The book is abundantly illustrated with all the various Courts, Warrens, Hutches, Fencing, etc., and also with excellent portraits of the most important species of rabbits throughout the world. 12mo ..... ............ 1.50

### Quinby's New Bee-Keeping.

The Mysteries of Bee-keeping Explained. Combining the results of Fifty Years' Experience, with the latest discoveries and inventions, and presenting the most approved methods, forming a complete work. Cloth, 12mo ............................................... 1.50

### Profits in Poultry.

Useful and Ornamental Breeds and their Profitable Management. This excellent work contains the combined experience of a number of practical men in all departments of poultry raising. It is profusely illustrated and forms an unique and important addition to our poultry literature. Cloth, 12mo ......... ...... .  - ....... ...... ... 1.00

### Barn Plans and Outbuildings.

Two Hundred and Fifty-seven Illustrations. A most Valuable Work, full of Ideas, Hints, Suggestions, Plans, etc., for the Construction of Barns and Outbuildings, by Practical writers Chapters are devoted, among other subjects, to the Economic Erection and Use of Barns. Grain Barns, Horse Barns, Cattle Barns, Sheep Barns, Corn Houses, Smoke Houses, Ice Houses, Pig Pens, Granaries, etc. There are likewise chapters upon Bird Houses, Dog Houses, Tool Sheds. Ventilator Roofs and Roofing, Doors and Fastenings, Work Shops, Poultry Houses, Manure Sheds, Barn Yards, Root Pits. etc. Recently published. Cloth, 12mo ............................................... 1.50

## Parsons on the Rose.

By Samuel B. Parsons. A treatise on the propagation, culture, and history of the rose. New and revised edition. In his work upon the rose, Mr. Parsons has gathered up the curious legends concerning the flower, and gives us an idea of the esteem in which it was held in former times. A simple garden classification has been adopted, and the leading varieties under each class enumerated and briefly described. The chapters on multiplication, cultivation, and training are very full, and the work is altogether one of the most complete before the public. Illustrated. Cloth, 12mo....................1.00

## Heinrich's Window Flower Garden.

The author is a practical florist, and this enterprising volume embodies his personal experiences in Window Gardening during a long period. New and enlarged edition. By Julius J. Heinrich. Fully Illustrated. Cloth, 12mo............................................... .75

## Liautard's Chart of the Age of the Domestic Animals.

Adopted by the United States Army. Enables one to accurately determine the age of horses, cattle, sheep, dogs, and pigs.......... .50

## Pedder's Land Measurer for Farmers.

A convenient Pocket Companion, showing at once the contents of any piece of land, when its length and width are known, up to 1,500 feet either way, with various other useful farm tables. Cloth, 18mo;
.60

## How to Plant and What to Do with the Crops.

With other valuable hints for the Farm, Garden and Orchard. By Mark W. Johnson. Illustrated. CONTENTS : Times for Sowing Seeds : Covering Seeds ; Field Crops ; Garden or Vegetable Seeds, Sweet Herbs, etc.; Tree Seeds ; Flower Seeds ; Fruit Trees : Distances Apart for Fruit Trees and Shrubs ; Profitable Farming ; Green or Manuring Crops : Root Crops ; Forage Plants : What to do with the Crops ; The Rotation of Crops ; Varieties ; Paper Covers, post-paid.......... .50

## Your Plants.

Plain and Practical Directions for the Treatment of Tender and Hardy Plants in the House and in the Garden. By James Sheehan. The above title well describes the character of the work—" Plain and Practical." The author, a commercial florist and gardener, has endeavored, in this work, to answer the many questions asked by his customers, as to the proper treatment of plants. The book shows all through that its author is a practical man, and he writes as one with a large store of experience. The work better meets the wants of the amateur who grows a few plants in the window, or has a small flower Garden, than a larger treatise intended for those who cultivate plants upon a more extended-scale. Price, post-paid, paper covers..................... .40

## Husmann's American Grape-Growing and Wine-Making.

By George Husmann of Talcoa vineyards, Napa, California. New and enlarged edition. With contributions from well-known grape-growers, giving a wide range of experience. The author of this book is a recognized authority on the subject. Cloth, 12mo............... 1.50

## The Scientific Angler.

A general and instructive work on Artistic Angling, by the late David Foster. Compiled by his Sons. With an Introductory Chapter and Copious Foot Notes, by William C. Harris, Editor of the "American Angler." Cloth, 12mo..................................................... 1.50

### Keeping One Cow.

A collection of Prize Essays, and selections from a number of other Essays, with editorial notes, suggestions, etc. This book gives the latest information, and in a clear and condensed form, upon the management of a single Milch Cow. Illustrated with full-page engravings of the most famous dairy cows. Recently published. Cloth, 12mo ............................................................ 1.00

### Law's Veterinary Adviser

A Guide to the Prevention and Treatment of Disease in Domestic Animals. This is one of the best works on this subject, and is especially designed to supply the need of the busy American Farmer, who can rarely avail himself of the advice of a Scientific Veterinarian. It is brought up to date and treats of the Prevention of Disease, as well as of the Remedies. By Prof. Jas. Law. Cloth, Crown 8vo..... 3.00

### Guenon's Treatise on Milch Cows.

A Treatise on the Bovine Species in General. An entirely new translation of the last edition of this popular and instructive book. By Thos. J. Hand, Secretary of the American Jersey Cattle Club. With over 100 Illustrations especially engraved for this work. Cloth, 12mo.
1.00

### The Cider Maker's Handbook.

A complete guide for making and keeping pure cider. By J. M. Trowbridge. Fully Illustrated. Cloth, 12mo...................... 1.00

### Long's Ornamental Gardening for Americans.

A treatise on Beautifying Homes, Rural Districts, and Cemeteries. A plain and practical work at a moderate price, with numerous illustrations, and instructions so plain that they may be readily followed. By Elias A. Long. Landscape Architect. Illustrated. Cloth, 12mo.
2.00

### The Dogs of Great Britain, America and Other Countries.

New, enlarged and revised edition. Their breeding, training and management, in health and disease; comprising all the essential parts of the two standard works on the dog, by "Stonehenge," thereby furnishing for $2 what once cost $11.25. Contains Lists of all Premiums given at the last Dog Shows. It Describes the Best Game and Hunting Grounds in America. Contains over One Hundred Beautiful Engravings, embracing most noted Dogs in both Continents, making together, with Chapters by American Writers, the most Complete Dog Book ever published. Cloth, 12mo............ ... .... ........ 2.00

### Stewart's Feeding Animals.

By Elliot W. Stewart. A new and valuable practical work upon the laws of animal growth, specially applied to the rearing and feeding horses, cattle, diary cows, sheep and swine. Illustrated. Cloth, 12mo
2.00

### How to Co-operate.

A Manual for Co-operators. By Herbert Myrick. This book describes the how rather than the wherefore of co-operation. In other words it tells how to manage a co-operative store, farm or factory, and co-operative dairying, banking and fire insurance, and co-operative farmers' and women's exchanges for both buying and selling. The directions given are based on the actual experience of successful co-operative enterprises in all parts of the United States. The character and usefulness of the book commend it to the attention of all men and women who desire to better their condition. 12mo. Cloth.............. 1.50

www.ingramcontent.com/pod-product-compliance
Lightning Source LLC
LaVergne TN
LVHW012203040326
832903LV00003B/97